T0291667

CAMBRIDGE LIBRARY COLLECTION

Books of enduring scholarly value

Darwin

Two hundred years after his birth and 150 years after the publication of 'On the Origin of Species', Charles Darwin and his theories are still the focus of worldwide attention. This series offers not only works by Darwin, but also the writings of his mentors in Cambridge and elsewhere, and a survey of the impassioned scientific, philosophical and theological debates sparked by his 'dangerous idea'.

The Life and Letters of Charles Darwin

This three-volume life of Charles Darwin, published five years after his death, was edited by his son Francis, who was his father's collaborator in experiments in botany and who after his death took on the responsibility of overseeing the publication of his remaining manuscript works and letters. In the preface to the first volume, Francis Darwin explains his editorial principles: 'In choosing letters for publication I have been largely guided by the wish to illustrate my father's personal character. But his life was so essentially one of work, that a history of the man could not be written without following closely the career of the author.' Among the family history, anecdotes and reminiscences of scientific colleagues is a short autobiographical essay which Charles Darwin wrote for his children and grandchildren, rather than for publication. This account of Darwin the man has never been bettered.

Cambridge University Press has long been a pioneer in the reissuing of out-of-print titles from its own backlist, producing digital reprints of books that are still sought after by scholars and students but could not be reprinted economically using traditional technology. The Cambridge Library Collection extends this activity to a wider range of books which are still of importance to researchers and professionals, either for the source material they contain, or as landmarks in the history of their academic discipline.

Drawing from the world-renowned collections in the Cambridge University Library, and guided by the advice of experts in each subject area, Cambridge University Press is using state-of-the-art scanning machines in its own Printing House to capture the content of each book selected for inclusion. The files are processed to give a consistently clear, crisp image, and the books finished to the high quality standard for which the Press is recognised around the world. The latest print-on-demand technology ensures that the books will remain available indefinitely, and that orders for single or multiple copies can quickly be supplied.

The Cambridge Library Collection will bring back to life books of enduring scholarly value across a wide range of disciplines in the humanities and social sciences and in science and technology.

The Life and Letters of Charles Darwin

Including an Autobiographical Chapter

VOLUME 1

CHARLES DARWIN
EDITED BY FRANCIS DARWIN

CAMBRIDGE
UNIVERSITY PRESS

CAMBRIDGE UNIVERSITY PRESS

Cambridge New York Melbourne Madrid Cape Town Singapore São Paolo Delhi

Published in the United States of America by Cambridge University Press, New York

www.cambridge.org
Information on this title: www.cambridge.org/9781108003445

© in this compilation Cambridge University Press 2009

This edition first published 1887
This digitally printed version 2009

ISBN 978-1-108-00344-5

FROM A PHOTOGRAPH (1854?) BY MESSRS. MAULL AND FOX.   ENGRAVED FOR
'HARPER'S MAGAZINE,' OCTOBER 1884.

Frontispiece, Vol. I.

THE

LIFE AND LETTERS

OF

CHARLES DARWIN,

INCLUDING

AN AUTOBIOGRAPHICAL CHAPTER.

EDITED BY HIS SON,

FRANCIS DARWIN.

IN THREE VOLUMES:—VOL. I.

LONDON:

JOHN MURRAY, ALBEMARLE STREET.

1887.

PREFACE.

———◆◆———

IN choosing letters for publication I have been largely guided by the wish to illustrate my father's personal character. But his life was so essentially one of work, that a history of the man could not be written without following closely the career of the author. Thus it comes about that the chief part of the book falls into chapters whose titles correspond to the names of his books.

In arranging the letters I have adhered as far as possible to chronological sequence, but the character and variety of his researches make a strictly chronological order an impossibility. It was his habit to work more or less simultaneously at several subjects. Experimental work was often carried on as a refreshment or variety, while books entailing reasoning and the marshalling of large bodies of facts were being written. Moreover, many of his researches were allowed to drop, and only resumed after an interval of years. Thus a rigidly chronological series of letters would present a patchwork of subjects, each of which would be difficult to follow. The Table of Contents will show in what way I have attempted to avoid this result. It will be seen, for instance, that the second volume is not chronologically continuous with the first. Again, in the third volume, the botanical work, which principally occupied my father during the later years of his life, is treated in a separate series of chapters.

In printing the letters I have followed (except in a few cases) the usual plan of indicating the existence of omissions or insertions. My father's letters give frequent evidence of having been written when he was tired or hurried. In a letter to a friend, or to one of his family, he frequently omitted the articles : these have been inserted without the usual indications, except in a few instances (*e.g.* Vol. I. p. 203), where it is of special interest to preserve intact the hurried character of the letter. Other small words, such as *of*, *to*, &c., have been inserted, usually within brackets. My father underlined many words in his letters; these have not always been given in italics,—a rendering which would have unfairly exaggerated their effect. I have not followed the originals as regards the spelling of names, the use of capital letters, or in the matter of punctuation.

The Diary or Pocket-book, from which quotations occur in the following pages, has been of value as supplying a framework of facts round which letters may be grouped. It is unfortunately written with great brevity, the history of a year being compressed into a page or less, and contains little more than the dates of the principal events of his life, together with entries as to his work, and as to the duration of his more serious illnesses. He rarely dated his letters, so that but for the Diary it would have been all but impossible to unravel the history of his books. It has also enabled me to assign dates to many letters which would otherwise have been shorn of half their value.

Of letters addressed to my father I have not made much use. It was his custom to file all letters received, and when his slender stock of files ("spits" as he called them) was exhausted, he would burn the letters of several years, in order that he might make use of the liberated "spits." This process, carried on for years, destroyed nearly all letters received before 1862. After that date he was persuaded to keep the more interesting letters, and these are preserved in an accessible form.

I have attempted to give, in Chapter III., some account of his manner of working. During the last eight years of his life I acted as his assistant, and thus had an opportunity of knowing something of his habits and methods.

I have received much help from my friends in the course of my work. To some I am indebted for reminiscences of my father, to others for information, criticisms, and advice. To all these kind coadjutors I gladly acknowledge my indebtedness. The names of some occur in connection with their contributions, but I do not name those to whom I am indebted for criticisms or corrections, because I should wish to bear alone the load of my short-comings, rather than to let any of it fall on those who have done their best to lighten it.

It will be seen how largely I am indebted to Sir Joseph Hooker for the means of illustrating my father's life. The readers of these pages will, I think, be grateful to Sir Joseph for the care with which he has preserved his valuable collection of letters, and I should wish to add my acknowledgment of the generosity with which he has placed it at my disposal, and for the kindly encouragement given throughout my work.

To Mr. Huxley I owe a debt of thanks, not only for much kind help, but for his willing compliance with my request that he should contribute a chapter on the reception of the 'Origin of Species.'

Finally, it is a pleasure to acknowledge the courtesy of the publishers of the 'Century Magazine' and of 'Harper's Magazine,' who have freely given me the use of their illustrations. To Messrs. Maull and Fox and Messrs. Elliott and Fry I am also indebted for their kindness in allowing me the use of reproductions of their photographs.

FRANCIS DARWIN.

CAMBRIDGE,
October, 1887.

TABLE OF CONTENTS.

VOLUME I.

VOLUME II.

VOLUME III.

BOTANICAL LETTERS.

APPENDICES.

ILLUSTRATIONS.

LIFE AND LETTERS

OF

CHARLES DARWIN.

—◦—

CHAPTER I.

THE DARWIN FAMILY.

THE earliest records of the family show the Darwins to have been substantial yeomen residing on the northern borders of Lincolnshire, close to Yorkshire. The name is now very unusual in England, but I believe that it is not unknown in the neighbourhood of Sheffield and in Lancashire. Down to the year 1600 we find the name spelt in a variety of ways —Derwent, Darwen, Darwynne, &c. It is possible, therefore, that the family migrated at some unknown date from York-shire, Cumberland, or Derbyshire, where Derwent occurs as the name of a river.

The first ancestor of whom we know was one William Darwin, who lived, about the year 1500, at Marton, near Gainsborough. His great grandson, Richard Darwyn, in-herited land at Marton and elsewhere, and in his will, dated 1584, "bequeathed the sum of 3s. 4d. towards the settynge up of the Queene's Majestie's armes over the quearie (choir) doore in the parishe churche of Marton." *

The son of this Richard, named William Darwin, and described as "gentleman," appears to have been a successful

---

* We owe a knowledge of these earlier members of the family to researches amongst the wills at Lincoln, made by the well-known genealogist, Colonel Chester.

man. Whilst retaining his ancestral land at Marton, he
acquired through his wife and by purchase an estate at
Cleatham, in the parish of Manton, near Kirton Lindsey, and
fixed his residence there. This estate remained in the family
down to the year 1760. A cottage with thick walls, some
fish-ponds and old trees, now alone show where the " Old
Hall" once stood, and a field is still locally known as the
" Darwin Charity," from being subject to a charge in favour
of the poor of Marton. William Darwin must, at least in part,
have owed his rise in station to his appointment in 1613 by
James I. to the post of Yeoman of the Royal Armoury of
Greenwich. The office appears to have been worth only £33
a year, and the duties were probably almost nominal ; he
held the post down to his death during the Civil Wars.

The fact that this William was a royal servant may explain
why his son, also named William, served when almost a boy
for the King, as "Captain-Lieutenant" in Sir William Pel-
ham's troop of horse. On the partial dispersion of the royal
armies, and the retreat of the remainder to Scotland, the boy's
estates were sequestrated by the Parliament, but they were
redeemed on his signing the Solemn League and Covenant,
and on his paying a fine which must have struck his finances
severely ; for in a petition to Charles II. he speaks of his
almost utter ruin from having adhered to the royal cause.

During the Commonwealth, William Darwin became a
barrister of Lincoln's Inn, and this circumstance probably led
to his marriage with the daughter of Erasmus Earle, serjeant-
at-law ; hence his great-grandson, Erasmus Darwin, the Poet,
derived his Christian name. He ultimately became Recorder
of the city of Lincoln.

The eldest son of the Recorder, again called William, was
born in 1655, and married the heiress of Robert Waring, a
member of a good Staffordshire family. This lady inherited
from the family of Lassells, or Lascelles, the manor and hall
of Elston, near Newark, which has remained ever since in the

family.* A portrait of this William Darwin at Elston shows him as a good-looking young man in a full-bottomed wig.

This third William had two sons, William, and Robert who was educated as a barrister. The Cleatham property was left to William, but on the termination of his line in daughters reverted to the younger brother, who had received Elston. On his mother's death Robert gave up his profession and resided ever afterwards at Elston Hall. Of this Robert, Charles Darwin writes †:—

" He seems to have had some taste for science, for he was an early member of the well-known Spalding Club ; and the celebrated antiquary Dr. Stukeley, in ' An Account of the almost entire Sceleton of a large Animal,' &c., published in the ' Philosophical Transactions,' April and May 1719, begins the paper as follows : ' Having an account from my friend, Robert Darwin, Esq., of Lincoln's Inn, a person of curiosity, of a human sceleton impressed in stone, found lately by the rector of Elston,' &c. Stukeley then speaks of it as a great rarity, 'the like whereof has not been observed before in this island to my knowledge.' Judging from a sort of litany written by Robert, and handed down in the family, he was a strong advocate of temperance, which his son ever afterwards so strongly advocated :—

> From a morning that doth shine,
> From a boy that drinketh wine,
> From a wife that talketh Latine,
> Good Lord deliver me !

---

* Captain Lassells, or Lascelles, of Elston was military secretary to Monk, Duke of Albemarle, during the Civil Wars. A large volume of account-books, countersigned in many places by Monk, are now in the possession of my cousin Francis Darwin. The accounts might possibly prove of interest to the antiquarian or historian. A portrait of Captain Lassells in armour, although used at one time as an archery-target by some small boys of our name, was not irretrievably ruined.

† What follows is quoted from Charles Darwin's biography of his grandfather, forming the preliminary notice to Ernst Krause's interesting essay, ' Erasmus Darwin,' London, 1879, p. 4.

" It is suspected that the third line may be accounted for by his wife, the mother of Erasmus, having been a very learned lady. The eldest son of Robert, christened Robert Waring, succeeded to the estate of Elston, and died there at the age of ninety-two, a bachelor. He had a strong taste for poetry, like his youngest brother Erasmus. Robert also cultivated botany, and, when an oldish man, he published his ' Principia Botanica.' This book in MS. was beautifully written, and my father [Dr. R. W. Darwin] declared that he believed it was published because his old uncle could not endure that such fine caligraphy should be wasted. But this was hardly just, as the work contains many curious notes on biology — a subject wholly neglected in England in the last century. The public, moreover, appreciated the book, as the copy in my possession is the third edition."

The second son, William Alvey, inherited Elston, and transmitted it to his granddaughter, the late Mrs. Darwin, of Elston and Creskeld. A third son, John, became rector of Elston, the living being in the gift of the family. The fourth son, and youngest child, was Erasmus Darwin, the poet and philosopher.

The table on page 5 shows Charles Darwin's descent from Robert, and his relationship to some other members of the family, whose names occur in his correspondence. Among these are included William Darwin Fox, one of his earliest correspondents, and Francis Galton, with whom he maintained a warm friendship for many years. Here also occurs the name of Francis Sacheverel Darwin, who inherited a love of natural history from Erasmus, and transmitted it to his son Edward Darwin, author (under the name of " High Elms ") of a 'Gamekeeper's Manual' (4th Edit. 1863), which shows keen observation of the habits of various animals.

It is always interesting to see how far a man's personal characteristics can be traced in his forefathers. Charles Darwin inherited the tall stature, but not the bulky figure of

TABLE OF RELATIONSHIP.

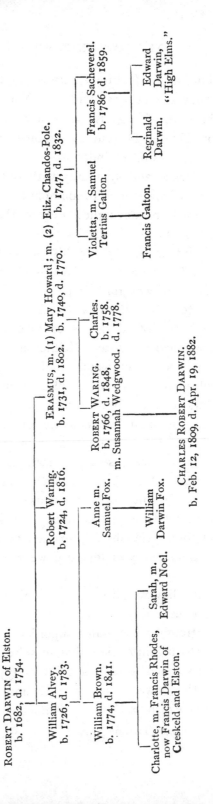

Erasmus ; but in his features there is no traceable resemblance to those of his grandfather. Nor, it appears, had Erasmus the love of exercise and of field-sports, so characteristic of Charles Darwin as a young man, though he had, like his grandson, an indomitable love of hard mental work. Benevolence and sympathy with others, and a great personal charm of manner, were common to the two. Charles Darwin possessed, in the highest degree, that "vividness of imagination" of which he speaks as strongly characteristic of Erasmus, and as leading "to his overpowering tendency to theorise and generalise." This tendency, in the case of Charles Darwin, was fully kept in check by the determination to test his theories to the utmost. Erasmus had a strong love of all kinds of mechanism, for which Charles Darwin had no taste. Neither had Charles Darwin the literary temperament which made Erasmus a poet as well as a philosopher. He writes of Erasmus : * "Throughout his letters I have been struck with his indifference to fame, and the complete absence of all signs of any over-estimation of his own abilities, or of the success of his works." These, indeed, seem indications of traits most strikingly prominent in his own character. Yet we get no evidence in Erasmus of the intense modesty and simplicity that marked Charles Darwin's whole nature. But by the quick bursts of anger provoked in Erasmus, at the sight of any inhumanity or injustice, we are again reminded of him.

On the whole, however, it seems to me that we do not know enough of the essential personal tone of Erasmus Darwin's character to attempt more than a superficial comparison ; and I am left with an impression that, in spite of many resemblances, the two men were of a different type. It has been shown that Miss Seward and Mrs. Schimmelpenninck have misrepresented Erasmus Darwin's character.† It is, however,

* 'Life of Erasmus Darwin,' p. 68.   † Ibid. pp. 77, 79, &c.

extremely probable that the faults which they exaggerate were to some extent characteristic of the man ; and this leads me to think that Erasmus had a certain acerbity or severity of temper which did not exist in his grandson.

The sons of Erasmus Darwin inherited in some degree his intellectual tastes, for Charles Darwin writes of them as follows* :—

"His eldest son, Charles (born September 3, 1758), was a young man of extraordinary promise, but died (May 15, 1778) before he was twenty-one years old, from the effects of a wound received whilst dissecting the brain of a child. He inherited from his father a strong taste for various branches of science, for writing verses, and for mechanics . . . He also inherited stammering. With the hope of curing him, his father sent him to France, when about eight years old (1766–67), with a private tutor, thinking that if he was not allowed to speak English for a time, the habit of stammering might be lost ; and it is a curious fact, that in after years, when speaking French, he never stammered. At a very early age he collected specimens of all kinds. When sixteen years old he was sent for a year to [Christ Church] Oxford, but he did not like the place, and thought (in the words of his father) that the 'vigour of his mind languished in the pursuit of classical elegance like Hercules at the distaff, and sighed to be removed to the robuster exercise of the medical school of Edinburgh.' He stayed three years at Edinburgh, working hard at his medical studies, and attending 'with diligence all the sick poor of the parish of Waterleith, and supplying them with the necessary medicines.' The Æsculapian Society awarded him its first gold medal for an experimental inquiry on pus and mucus. Notices of him appeared in various journals ; and all the writers agree about his uncommon energy and abilities. He seems like his father to have excited the warm affection of his friends. Professor

* 'Life of Erasmus Darwin,' p. 80.

Andrew Duncan . . . . spoke . . . . about him with the
warmest affection forty-seven years after his death when
I was a young medical student at Edinburgh . . .

"About the character of his second son Erasmus (born 1759),
I have little to say, for though he wrote poetry, he seems to
have had none of the other tastes of his father. He had,
however, his own peculiar tastes, viz. genealogy, the collecting
of coins, and statistics. When a boy he counted all the
houses in the city of Lichfield, and found out the number of
inhabitants in as many as he could; he thus made a census,
and when a real one was first made, his estimate was found to
be nearly accurate. His disposition was quiet and retiring.
My father had a very high opinion of his abilities, and this
was probably just, for he would not otherwise have been
invited to travel with, and pay long visits to, men so distinguished in different ways as Boulton the engineer, and Day
the moralist and novelist." His death by suicide, in 1799,
seems to have taken place in a state of incipient insanity.

Robert Waring, the father of Charles Darwin, was born
May 30, 1766, and entered the medical profession like his
father. He studied for a few months at Leyden, and took
his M.D.* at that University on Feb. 26, 1785. "His father"
(Erasmus) "brought † him to Shrewsbury before he was
twenty-one years old (1787), and left him £20, saying, 'Let
me know when you want more, and I will send it you.' His

---

* I owe this information to the
kindness of Professor Rauwenhoff,
Director of the Archives at Leyden.
He quotes from the catalogue of
doctors that "Robertus Waring
Darwin, Anglo-britannus,"defended
(Feb. 26, 1785) in the Senate a
Dissertation on the coloured images
seen after looking at a bright object,
and "Medicinæ Doctor creatus est
a clar. Paradijs." The archives of
Leyden University are so complete

that Professor Rauwenhoff is able
to tell me that my grandfather lived
together with a certain "Petrus
Crompton, Anglus," in lodgings in
the Apothekersdijk. Dr. Darwin's
Leyden dissertation was published
in the 'Philosophical Transactions,'
and my father used to say that the
work was in fact due to Erasmus
Darwin.—F. D.

† 'Life of Erasmus Darwin,'
p. 85.

uncle, the rector of Elston, afterwards also sent him £20, and this was the sole pecuniary aid * which he ever received . . . Erasmus tells Mr. Edgeworth that his son Robert, after being settled in Shrewsbury for only six months, 'already had between forty and fifty patients.' By the second year he was in considerable, and ever afterwards in very large, practice."

Robert Waring Darwin married (April 18, 1796) Susannah, the daughter of his father's friend, Josiah Wedgwood, of Etruria, then in her thirty-second year. We have a miniature of her, with a remarkably sweet and happy face, bearing some resemblance to the portrait by Sir Joshua Reynolds of her father; a countenance expressive of the gentle and sympathetic nature which Miss Meteyard ascribes to her.† She died July 15, 1817, thirty-two years before her husband, whose death occurred on November 13, 1848. Dr. Darwin lived before his marriage for two or three years on St. John's Hill, afterwards at the Crescent, where his eldest daughter Marianne was born, lastly at the "Mount," in the part of Shrewsbury known as Frankwell, where the other children were born. This house was built by Dr. Darwin about 1800, it is now in the possession of Mr. Spencer Phillips, and has undergone but little alteration. It is a large, plain, square, red-brick house, of which the most attractive feature is the pretty green-house, opening out of the morning-room.

The house is charmingly placed, on the top of a steep bank leading down to the Severn. The terraced bank is traversed by a long walk, leading from end to end, still called "the Doctor's Walk." At one point in this walk grows a Spanish chestnut, the branches of which bend back parallel to themselves in a curious manner, and this was Charles Darwin's favourite tree as a boy, where he and his sister Catherine had each their special seat.

The Doctor took great pleasure in his garden, planting it

* See errata.   † 'A Group of Englishmen,' by Miss Meteyard, 1871.

with ornamental trees and shrubs, and being especially successful in fruit-trees ; and this love of plants was, I think, the
only taste kindred to natural history which he possessed. Of
the "Mount pigeons," which Miss Meteyard describes as
illustrating Dr. Darwin's natural-history tastes, I have not
been able to hear from those most capable of knowing. Miss
Meteyard's account of him is not quite accurate in a few
points. For instance, it is incorrect to describe Dr. Darwin as
having a philosophical mind ; his was a mind especially given
to detail, and not to generalising. Again, those who knew him
intimately describe him as eating remarkably little, so that
he was not "a great feeder, eating a goose for his dinner, as
easily as other men do a partridge." * In the matter of dress
he was conservative, and wore to the end of his life knee-
breeches and drab gaiters, which, however, certainly did not,
as Miss Meteyard says, button above the knee—a form of
costume chiefly known to us in grenadiers of Queen Anne's
day, and in modern wood-cutters and ploughboys.

Charles Darwin had the strongest feeling of love and
respect for his father's memory. His recollection of everything that was connected with him was peculiarly distinct,
and he spoke of him frequently ; generally prefacing an anecdote with some such phrase as, "My father, who was the
wisest man I ever knew, &c." It was astonishing how clearly
he remembered his father's opinions, so that he was able to
quote some maxims or hint of his in most cases of illness.
As a rule he put small faith in doctors, and thus his
unlimited belief in Dr. Darwin's medical instinct, and
methods of treatment was all the more striking.

His reverence for him was boundless and most touching. He
would have wished to judge everything else in the world dis-
passionately, but anything his father had said was received
with almost implicit faith. His daughter Mrs. Litchfield
remembers him saying that he hoped none of his sons would

* 'A Group of Englishmen,' p. 263.

ever believe anything because he said it, unless they were themselves convinced of its truth,—a feeling in striking contrast with his own manner of faith.

A visit which Charles Darwin made to Shrewsbury in 1869 left on the mind of his daughter who accompanied him a strong impression of his love for his old home. The then tenant of the Mount showed them over the house, &c., and with mistaken hospitality remained with the party during the whole visit. As they were leaving, Charles Darwin said, with a pathetic look of regret, "If I could have been left alone in that green-house for five minutes, I know I should have been able to see my father in his wheel-chair as vividly as if he had been there before me."

Perhaps this incident shows what I think is the truth, that the memory of his father he loved the best, was that of him as an old man. Mrs. Litchfield has noted down a few words which illustrate well his feeling towards his father. She describes him as saying with the most tender respect, "I think my father was a little unjust to me when I was young, but afterwards I am thankful to think I became a prime favourite with him." She has a vivid recollection of the expression of happy reverie that accompanied these words, as if he were reviewing the whole relation, and the remembrance left a deep sense of peace and gratitude.

What follows was added by Charles Darwin to his autobiographical ' Recollections,' and was written about 1877 or 1878.

"I may here add a few pages about my father, who was in many ways a remarkable man.

"He was about 6 feet 2 inches in height, with broad shoulders, and very corpulent, so that he was the largest man whom I ever saw. When he last weighed himself, he was 24 stone, but afterwards increased much in weight. His chief mental characteristics were his powers of obser-

vation and his sympathy, neither of which have I ever seen
exceeded or even equalled. His sympathy was not only
with the distresses of others, but in a greater degree with
the pleasures of all around him. This led him to be always
scheming to give pleasure to others, and, though hating
extravagance, to perform many generous actions. For
instance, Mr. B——, a small manufacturer in Shrewsbury,
came to him one day, and said he should be bankrupt unless
he could at once borrow £10,000, but that he was unable
to give any legal security. My father heard his reasons
for believing that he could ultimately repay the money, and
from [his] intuitive perception of character felt sure that he
was to be trusted. So he advanced this sum, which was a
very large one for him while young, and was after a time
repaid.

"I suppose that it was his sympathy which gave him unbounded power of winning confidence, and as a consequence
made him highly successful as a physician. He began to
practise before he was twenty-one years old, and his fees
during the first year paid for the keep of two horses and a
servant. On the following year his practice was large, and so
continued for about sixty years, when he ceased to attend on
any one. His great success as a doctor was the more remarkable, as he told me that he at first hated his profession so
much that if he had been sure of the smallest pittance, or if
his father had given him any choice, nothing should have
induced him to follow it. To the end of 'his life, the thought
of an operation almost sickened him, and he could scarcely
endure to see a person bled—a horror which he has transmitted to me—and I remember the horror which I felt as a
schoolboy in reading about Pliny (I think) bleeding to death
in a warm bath. . . .

"Owing to my father's power of winning confidence, many
patients, especially ladies, consulted him when suffering from
any misery, as a sort of Father-Confessor. He told me that

they always began by complaining in a vague manner about their health, and by practice he soon guessed what was really the matter. He then suggested that they had been suffering in their minds, and now they would pour out their troubles, and he heard nothing more about the body. . . . Owing to my father's skill in winning confidence he received many strange confessions of misery and guilt. He often remarked how many miserable wives he had known. In several instances husbands and wives had gone on pretty well together for between twenty and thirty years, and then hated each other bitterly; this he attributed to their having lost a common bond in their young children having grown up.

"But the most remarkable power which my father possessed was that of reading the characters, and even the thoughts of those whom he saw even for a short time. We had many instances of the power, some of which seemed almost supernatural. It saved my father from ever making (with one exception, and the character of this man was soon discovered) an unworthy friend. A strange clergyman came to Shrewsbury, and seemed to be a rich man; everybody called on him, and he was invited to many houses. My father called, and on his return home told my sisters on no account to invite him or his family to our house; for he felt sure that the man was not to be trusted. After a few months he suddenly bolted, being heavily in debt, and was found out to be little better than an habitual swindler. Here is a case of trustfulness which not many men would have ventured on. An Irish gentleman, a complete stranger, called on my father one day, and said that he had lost his purse, and that it would be a serious inconvenience to him to wait in Shrewsbury until he could receive a remittance from Ireland. He then asked my father to lend him £20, which was immediately done, as my father felt certain that the story was a true one. As soon as a letter could arrive from Ireland, one came with the most profuse thanks, and enclosing, as he said, a £20 Bank of England note, but no note was enclosed. I asked my father whether this did not stagger him, but he answered 'not in the least.' On the next day another letter came with many apologies for having forgotten (like a true Irishman) to put the note into his letter of the day before. . . . [A gentleman] brought his nephew, who was insane but quite gentle, to my father ; and the young man's insanity led him to accuse himself of all the crimes under heaven. When my father afterwards talked over the matter with the uncle, he said, 'I am sure that your nephew is really guilty of . . . a heinous crime.' Whereupon [the gentleman] said, 'Good God, Dr. Darwin, who told you ; we thought that no human being knew the fact except ourselves!' My father told me the story many years after the event, and I asked him how he distinguished the true from the false self-accusations ; and it was very characteristic of my father that he said he could not explain how it was.

"The following story shows what good guesses my father could make. Lord Shelburne, afterwards the first Marquis of Lansdowne, was famous (as Macaulay somewhere remarks) for his knowledge of the affairs of Europe, on which he greatly prided himself. He consulted my father medically, and afterwards harangued him on the state of Holland. My father had studied medicine at Leyden, and one day [while there] went a long walk into the country with a friend who took him to the house of a clergyman (we will say the Rev. Mr. A——, for I have forgotten his name), who had married an Englishwoman. My father was very hungry, and there was little for luncheon except cheese, which he could never eat. The old lady was surprised and grieved at this, and assured my father that it was an excellent cheese, and had been sent her from Bowood, the seat of Lord Shelburne. My father wondered why a cheese should be sent her from Bowood, but thought nothing more about it until it flashed across his mind many years afterwards, whilst Lord Shelburne was talking about

Holland. So he answered, 'I should think from what I saw of the Rev. Mr. A——, that he was a very able man, and well acquainted with the state of Holland.' My father saw that the Earl, who immediately changed the conversation, was much startled. On the next morning my father received a note from the Earl, saying that he had delayed starting on his journey, and wished particularly to see my father. When he called, the Earl said, 'Dr. Darwin, it is of the utmost importance to me and to the Rev. Mr. A—— to learn how you have discovered that he is the source of my information about Holland.' So my father had to explain the state of the case, and he supposed that Lord Shelburne was much struck with his diplomatic skill in guessing, for during many years afterwards he received many kind messages from him through various friends. I think that he must have told the story to his children ; for Sir C. Lyell asked me many years ago why the Marquis of Lansdowne (the son or grandson of the first marquis) felt so much interest about me, whom he had never seen, and my family. When forty new members (the forty thieves as they were then called) were added to the Athenæum Club, there was much canvassing to be one of them ; and without my having asked any one, Lord Lansdowne proposed me and got me elected. If I am right in my supposition, it was a queer concatenation of events that my father not eating cheese half-a-century before in Holland led to my election as a member of the Athenæum.

"The sharpness of his observation led him to predict with remarkable skill the course of any illness, and he suggested endless small details of relief. I was told that a young doctor in Shrewsbury, who disliked my father, used to say that he was wholly unscientific, but owned that his power of predicting the end of an illness was unparalleled. Formerly when he thought that I should be a doctor, he talked much to me about his patients. In the old days the practice of bleeding largely was universal, but my father maintained that far more evil was thus caused than good done ; and he advised me if ever I was myself ill not to allow any doctor to take more than an extremely small quantity of blood. Long before typhoid fever was recognised as distinct, my father told me that two utterly distinct kinds of illness were confounded under the name of typhus fever. He was vehement against drinking, and was convinced of both the direct and inherited evil effects of alcohol when habitually taken even in moderate quantity in a very large majority of cases. But he admitted and advanced instances of certain persons who could drink largely during their whole lives without apparently suffering any evil effects, and he believed that he could often beforehand tell who would thus not suffer. He himself never drank a drop of any alcoholic fluid. This remark reminds me of a case showing how a witness under the most favourable circumstances may be utterly mistaken. A gentleman-farmer was strongly urged by my father not to drink, and was encouraged by being told that he himself never touched any spirituous liquor. Whereupon the gentleman said, ' Come, come, Doctor, this won't do—though it is very kind of you to say so for my sake—for I know that you take a very large glass of hot gin and water every evening after your dinner.' * So my father asked him how he knew this. The man answered, ' My cook was your kitchen-maid for two or three years, and she saw the butler every day prepare and take to you the gin and water.' The explanation was that my father had the odd habit of drinking hot water in a very tall and large glass after his dinner ; and the butler used first to put some cold water in the glass, which the girl mistook for gin, and then filled it up with boiling water from the kitchen boiler.

" My father used to tell me many little things which he had found useful in his medical practice. Thus ladies often

* This belief still survives, and   1884 by an old inhabitant of
was mentioned to my brother in   Shrewsbury.—F. D.

cried much while telling him their troubles, and thus caused much loss of his precious time. He soon found that begging them to command and restrain themselves, always made them weep the more, so that afterwards he always encouraged them to go on crying, saying that this would relieve them more than anything else, and with the invariable result that they soon ceased to cry, and he could hear what they had to say and give his advice. When patients who were very ill craved for some strange and unnatural food, my father asked them what had put such an idea into their heads : if they answered that they did not know, he would allow them to try the food, and often with success, as he trusted to their having a kind of instinctive desire ; but if they answered that they had heard that the food in question had done good to some one else, he firmly refused his assent.

"He gave one day an odd little specimen of human nature. When a very young man he was called in to consult with the family physician in the case of a gentleman of much distinction in Shropshire. The old doctor told the wife that the illness was of such a nature that it must end fatally. My father took a different view and maintained that the gentleman would recover : he was proved quite wrong in all respects (I think by autopsy) and he owned his error. He was then convinced that he should never again be consulted by this family ; but after a few months the widow sent for him, having dismissed the old family doctor. My father was so much surprised at this, that he asked a friend of the widow to find out why he was again consulted. The widow answered her friend, that 'she would never again see the odious old doctor who said from the first that her husband would die, while Dr. Darwin always maintained that he would recover !' In another case my father told a lady that her husband would certainly die. Some months afterwards he saw the widow who was a very sensible woman, and she said, 'You are a very young man, and allow me to advise you always to give as

long as you possibly can, hope to any near relative nursing a
patient. You made me despair, and from that moment I lost
strength.' My father said that he had often since seen the
paramount importance, for the sake of the patient, of keeping
up the hope and with it the strength of the nurse in charge.
This he sometimes found difficult to do compatibly with
truth. One old gentleman, however, caused him no such
perplexity. He was sent for by Mr. P——, who said, 'From
all that I have seen and heard of you I believe that you are
the sort of man who will speak the truth, and if I ask, you
will tell me when I am dying. Now I much desire that you
should attend me, if you will promise, whatever I may say,
always to declare that I am not going to die.' My father
acquiesced on the understanding that his words should in fact
have no meaning.

" My father possessed an extraordinary memory, especially
for dates, so that he knew, when he was very old, the day of
the birth, marriage, and death of a multitude of persons in
Shropshire ; and he once told me that this power annoyed
him ; for if he once heard a date, he could not forget it ; and
thus the deaths of many friends were often recalled to his
mind. Owing to his strong memory he knew an extraordinary
number of curious stories, which he liked to tell, as he was a
great talker. He was generally in high spirits, and laughed
and joked with every one—often with his servants—with the
utmost freedom ; yet he had the art of making every one obey
him to the letter. Many persons were much afraid of him. I
remember my father telling us one day, with a laugh, that several
persons had asked him whether Miss ——, a grand old lady in
Shropshire, had called on him, so that at last he enquired
why they asked him ; and was told that Miss ——, whom
my father had somehow mortally offended, was telling everybody that she would call and tell ' that fat old doctor very
plainly what she thought of him.' She had already called,
but her courage had failed, and no one could have been more

courteous and friendly. As a boy, I went to stay at the house of ——, whose wife was insane ; and the poor creature, as soon as she saw me, was in the most abject state of terror that I ever saw, weeping bitterly and asking me over and over again, ' Is your father coming? ' but was soon pacified. On my return home, I asked my father why she was so frightened, and he answered he was very glad to hear it, as he had frightened her on purpose, feeling sure that she would be kept in safety and much happier without any restraint, if her husband could influence her, whenever she became at all violent, by proposing to send for Dr. Darwin ; and these words succeeded perfectly during the rest of her long life.

" My father was very sensitive, so that many small events annoyed him or pained him much. I once asked him, when he was old and could not walk, why he did not drive out for exercise ; and he answered, ' Every road out of Shrewsbury is associated in my mind with some painful event.' Yet he was generally in high spirits. He was easily made very angry, but his kindness was unbounded. He was widely and deeply loved.

" He was a cautious and good man of business, so that he hardly ever lost money by any investment, and left to his children a very large property. I remember a story showing how easily utterly false beliefs originate and spread. Mr. E——, a squire of one of the oldest families in Shropshire, and head partner in a bank, committed suicide. My father was sent for as a matter of form, and found him dead. I may mention, by the way, to show how matters were managed in those old days, that because Mr. E—— was a rather great man, and universally respected, no inquest was held over his body. My father, in returning home, thought it proper to call at the bank (where he had an account) to tell the managing partners of the event, as it was not improbable that it would cause a run on the bank. Well, the story was spread far and wide, that my father went into the bank, drew out all his money, left the

C 2

bank, came back again, and said, 'I may just tell you that
Mr. E—— has killed himself,' and then departed. It seems
that it was then a common belief that money withdrawn from a
bank was not safe until the person had passed out through
the door of the bank. My father did not hear this story till some
little time afterwards, when the managing partner said that
he had departed from his invariable rule of never allowing any
one to see the account of another man, by having shown the
ledger with my father's account to several persons, as this
proved that my father had not drawn out a penny on that
day. It would have been dishonourable in my father to have
used his professional knowledge for his private advantage.
Nevertheless, the supposed act was greatly admired by some
persons ; and many years afterwards, a gentleman remarked,
' Ah, Doctor, what a splendid man of business you were in
so cleverly getting all your money safe out of that bank !'

"My father's mind was not scientific, and he did not try to
generalise his knowledge under general laws ; yet he formed a
theory for almost everything which occurred. I do not think
I gained much from him intellectually ; but his example
ought to have been of much moral service to all his children.
One of his golden rules (a hard one to follow) was, 'Never
become the friend of any one whom you cannot respect.'"

Dr. Darwin had six children :* Marianne, married Dr. Henry
Parker; Caroline, married Josiah Wedgwood; Erasmus Alvey ;
Susan, died unmarried ; Charles Robert ; Catherine, married
Rev. Charles Langton.

The elder son, Erasmus, was born in 1804, and died unmarried at the age of seventy-seven.

He, like his brother, was educated at Shrewsbury School
and at Christ's College, Cambridge. He studied medicine at
Edinburgh and in London, and took the degree of Bachelor
of Medicine at Cambridge. He never made any pretence of

* Of these Mrs. Wedgwood is now the sole survivor.

practising as a doctor, and, after leaving Cambridge, lived a quiet life in London.

There was something pathetic in Charles Darwin's affection for his brother Erasmus, as if he always recollected his solitary life, and the touching patience and sweetness of his nature. He often spoke of him as "Poor old Ras," or "Poor dear old Philos "—I imagine Philos (Philosopher) was a relic of the days when they worked at chemistry in the tool-house at Shrewsbury—a time of which he always preserved a pleasant memory. Erasmus being rather more than four years older than Charles Darwin, they were not long together at Cambridge, but previously at Edinburgh they lived in the same lodgings, and after the Voyage they lived for a time together in Erasmus' house in Great Marlborough Street. At this time also he often speaks with much affection of Erasmus in his letters to Fox, using words such as "my dear good old brother." In later years Erasmus Darwin came to Down occasionally, or joined his brother's family in a summer holiday. But gradually it came about that he could not, through ill health, make up his mind to leave London, and then they only saw each other when Charles Darwin went for a week at a time to his brother's house in Queen Anne Street.

The following note on his brother's character was written by Charles Darwin at about the same time that the sketch of his father was added to the 'Recollections':—

"My brother Erasmus possessed a remarkably clear mind with extensive and diversified tastes and knowledge in literature, art, and even in science. For a short time he collected and dried plants, and during a somewhat longer time experimented in chemistry. He was extremely agreeable, and his wit often reminded me of that in the letters and works of Charles Lamb. He was very kind-hearted. . . . His health from his boyhood had been weak, and as a consequence he failed in energy. His spirits were not high, sometimes low, more especially during early and middle manhood. He read much, even whilst a boy, and at school encouraged me to read, lending me books. Our minds and tastes were, however, so different, that I do not think I owe much to him intellectually. I am inclined to agree with Francis Galton in believing that education and environment produce only a small effect on the mind of any one, and that most of our qualities are innate."

Erasmus Darwin's name, though not known to the general public, may be remembered from the sketch of his character in Carlyle's ' Reminiscences,' which I here reproduce in part:—

" Erasmus Darwin, a most diverse kind of mortal, came to seek us out very soon ('had heard of Carlyle in Germany, &c.') and continues ever since to be a quiet house-friend, honestly attached ; though his visits latterly have been rarer and rarer, health so poor, I so occupied, &c., &c. He had something of original and sarcastically ingenious in him, one of the sincerest, naturally truest, and most modest of men ; elder brother of Charles Darwin (the famed Darwin on Species of these days) to whom I rather prefer him for intellect, had not his health quite doomed him to silence and patient idleness. . . . My dear one had a great favour for this honest Darwin always ; many a road, to shops and the like, he drove her in his cab (Darwingium Cabbum comparable to Georgium Sidus) in those early days when even the charge of omnibuses was a consideration, and his sparse utterances, sardonic often, were a great amusement to her. ' A perfect gentleman ' she at once discerned him to be, and of sound worth and kindliness in the most unaffected form." *

Charles Darwin did not appreciate this sketch of his brother;

* Carlyle's ' Reminiscences,' vol. ii. p. 208.

he thought Carlyle had missed the essence of his most lovable nature.

I am tempted by the wish of illustrating further the character of one so sincerely beloved by all Charles Darwin's children, to reproduce a letter to the *Spectator* (Sept. 3, 1881) by his cousin Miss Julia Wedgwood.

"A portrait from Mr. Carlyle's portfolio not regretted by any who loved the original, surely confers sufficient distinction to warrant a few words of notice, when the character it depicts is withdrawn from mortal gaze. Erasmus, the only brother of Charles Darwin, and the faithful and affectionate old friend of both the Carlyles, has left a circle of mourners who need no tribute from illustrious pen to embalm the memory so dear to their hearts; but a wider circle must have felt some interest excited by that tribute, and may receive with a certain attention the record of a unique and indelible impression, even though it be made only on the hearts of those who cannot bequeath it, and with whom, therefore, it must speedily pass away. They remember it with the same distinctness as they remember a creation of genius; it has in like manner enriched and sweetened life, formed a common meeting-point for those who had no other; and, in its strong fragrance of individuality, enforced that respect for the idiosyncrasies of human character without which moral judgment is always hard and shallow, and often unjust. Carlyle was one to find a peculiar enjoyment in the combination of liveliness and repose which gave his friend's society an influence at once stimulating and soothing, and the warmth of his appreciation was not made known first in its posthumous expression; his letters of anxiety nearly thirty years ago, when the frail life which has been prolonged to old age was threatened by serious illness, are still fresh in my memory. The friendship was equally warm with both husband and wife. I remember well a pathetic little remonstrance from her

elicited by an avowal from Erasmus Darwin, that he preferred
cats to dogs, which she felt a slur on her little 'Nero ;' and
the tones in which she said, 'Oh, but you are fond of dogs!
you are too kind not to be,' spoke of a long vista of small,
gracious kindnesses, remembered with a tender gratitude. He
was intimate also with a person whose friends, like those of
Mr. Carlyle, have not always had cause to congratulate themselves on their place in her gallery,—Harriet Martineau. I
have heard him more than once call her a faithful friend, and
it always seemed to me a curious tribute to something in the
friendship that he alone supplied ; but if she had written of
him at all, I believe the mention, in its heartiness of appreciation, would have afforded a rare and curious meeting-point
with the other 'Reminiscences,' so like and yet so unlike. It is
not possible to transfer the impression of a character ; we can
only suggest it by means of some resemblance ; and it is a
singular illustration of that irony which checks or directs our
sympathies, that in trying to give some notion of the man
whom, among those who were not his kindred, Carlyle appears
to have most loved, I can say nothing more descriptive than
that he seems to me to have had something in common with
the man whom Carlyle least appreciated. The society of
Erasmus Darwin had, to my mind, much the same charm as
the writings of Charles Lamb. There was the same kind of
playfulness, the same lightness of touch, the same tenderness,
perhaps the same limitations. On another side of his nature,
I have often been reminded of him by the quaint, delicate
humour, the superficial intolerance, the deep springs of pity,
the peculiar mixture of something pathetic with a sort of gay
scorn, entirely remote from contempt, which distinguish the
Ellesmere of Sir Arthur Helps' earlier dialogues. Perhaps
we recall such natures most distinctly, when such a resemblance
is all that is left of them. The character is not merged in the
creation ; and what we lose in the power to communicate our
impression, we seem to gain in its vividness. Erasmus Darwin

has passed away in old age, yet his memory retains something of a youthful fragrance ; his influence gave much happiness, of a kind usually associated with youth, to many lives besides the illustrious one whose records justify, though certainly they do not inspire, the wish to place this fading chaplet on his grave."

The foregoing pages give, in a fragmentary manner, as much perhaps as need be told of the family from which Charles Darwin came, and may serve as an introduction to the autobiographical chapter which follows.

CHAPTER II.

AUTOBIOGRAPHY.

[My father's autobiographical recollections, given in the present chapter, were written for his children,—and written without any thought that they would ever be published. To many this may seem an impossibility; but those who knew my father will understand how it was not only possible, but natural. The autobiography bears the heading, 'Recollections of the Development of my Mind and Character,' and end with the following note :—"Aug. 3, 1876. This sketch of my life was begun about May 28th at Hopedene,* and since then I have written for nearly an hour on most afternoons." It will easily be understood that, in a narrative of a personal and intimate kind written for his wife and children, passages should occur which must here be omitted ; and I have not thought it necessary to indicate where such omissions are made. It has been found necessary to make a few corrections of obvious verbal slips, but the number of such alterations has been kept down to the minimum.—F. D.]

A German Editor having written to me for an account of the development of my mind and character with some sketch of my autobiography, I have thought that the attempt would amuse me, and might possibly interest my children or their children. I know that it would have interested me greatly to have read even

* Mr. Hensleigh Wedgwood's house in Surrey.

so short and dull a sketch of the mind of my grandfather, written by himself, and what he thought and did, and how he worked. I have attempted to write the following account of myself, as if I were a dead man in another world looking back at my own life. Nor have I found this difficult, for life is nearly over with me. I have taken no pains about my style of writing.

I was born at Shrewsbury on February 12th, 1809, and my earliest recollection goes back only to when I was a few months over four years old, when we went to near Abergele for sea-bathing, and I recollect some events and places there with some little distinctness.

My mother died in July 1817, when I was a little over eight years old, and it is odd that I can remember hardly anything about her except her death-bed, her black velvet gown, and her curiously constructed work-table. In the spring of this same year I was sent to a day-school in Shrewsbury, where I stayed a year. I have been told that I was much slower in learning than my younger sister Catherine, and I believe that I was in many ways a naughty boy.

By the time I went to this day-school * my taste

* Kept by Rev. G. Case, minister of the Unitarian Chapel in the High Street. Mrs. Darwin was a Unitarian and attended Mr. Case's chapel, and my father as a little boy went there with his elder sisters. But both he and his brother were christened and intended to belong to the Church of England; and after his early boyhood he seems usually to have gone to church and not to Mr. Case's. It appears (*St. James' Gazette*, Dec. 15, 1883) that a mural tablet has been erected to his memory in the chapel, which is now known as the 'Free Christian Church.'— F. D.

for natural history, and more especially for collecting, was well developed. I tried to make out the names of plants,* and collected all sorts of things, shells, seals, franks, coins, and minerals. The passion for collecting which leads a man to be a systematic naturalist, a virtuoso, or a miser, was very strong in me, and was clearly innate, as none of my sisters or brother ever had this taste.

One little event during this year has fixed itself very firmly in my mind, and I hope that it has done so from my conscience having been afterwards sorely troubled by it; it is curious as showing that apparently I was interested at this early age in the variability of plants! I told another little boy (I believe it was Leighton, who afterwards became a well-known lichenologist and botanist), that I could produce variously coloured polyanthuses and primroses by watering them with certain coloured fluids, which was of course a monstrous fable, and had never been tried by me. I may here also confess that as a little boy I was much given to inventing deliberate falsehoods, and this was always done for the sake of causing excitement. For instance, I once gathered much valuable fruit from my father's trees and hid it in the shrubbery, and then ran in breathless

* Rev. W. A. Leighton, who was a schoolfellow of my father's at Mr. Case's school, remembers his bringing a flower to school and saying that his mother had taught him how by looking at the inside of the blossom the name of the plant could be discovered. Mr. Leighton goes on, "This greatly roused my attention and curiosity, and I inquired of him repeatedly how this could be done?"—but his lesson was naturally enough not transmissible. —F. D.

haste to spread the news that I had discovered a
hoard of stolen fruit.

I must have been a very simple little fellow when I
first went to the school. A boy of the name of
Garnett took me into a cake shop one day, and
bought some cakes for which he did not pay, as the
shopman trusted him. When we came out I asked
him why he did not pay for them, and he instantly
answered, " Why, do you not know that my uncle left
a great sum of money to the town on condition that
every tradesman should give whatever was wanted
without payment to any one who wore his old hat and
moved [it] in a particular manner ?" and he then showed
me how it was moved. He then went into another
shop where he was trusted, and asked for some small
article, moving his hat in the proper manner, and of
course obtained it without payment. When we came
out he said, "Now if you like to go by yourself into
that cake-shop (how well I remember its exact position) I will lend you my hat, and you can get whatever you like if you move the hat on your head
properly." I gladly accepted the generous offer, and
went in and asked for some cakes, moved the old hat
and was walking out of the shop, when the shopman
made a rush at me, so I dropped the cakes and ran
for dear life, and was astonished by being greeted
with shouts of laughter by my false friend Garnett.

I can say in my own favour that I was as a boy
humane, but I owed this entirely to the instruction
and example of my sisters. I doubt indeed whether
humanity is a natural or innate quality. I was very

fond of collecting eggs, but I never took more than a
single egg out of a bird's nest, except on one single
occasion, when I took all, not for their value, but from
a sort of bravado.

I had a strong taste for angling, and would sit for
any number of hours on the bank of a river or pond
watching the float; when at Maer * I was told that
I could kill the worms with salt and water, and from
that day I never spitted a living worm, though at the
expense probably of some loss of success.

Once as a very little boy whilst at the day school,
or before that time, I acted cruelly, for I beat a
puppy, I believe, simply from enjoying the sense of
power; but the beating could not have been severe,
for the puppy did not howl, of which I feel sure, as
the spot was near the house. This act lay heavily on
my conscience, as is shown by my remembering the
exact spot where the crime was committed. It probably lay all the heavier from my love of dogs being
then, and for a long time afterwards, a passion. Dogs
seemed to know this, for I was an adept in robbing
their love from their masters.

I remember clearly only one other incident during
this year whilst at Mr. Case's daily school,—namely,
the burial of a dragoon soldier; and it is surprising
how clearly I can still see the horse with the man's
empty boots and carbine suspended to the saddle, and
the firing over the grave. This scene deeply stirred
whatever poetic fancy there was in me.

In the summer of 1818 I went to Dr. Butler's great

* The house of his uncle, Josiah Wedgwood.

school in Shrewsbury, and remained there for seven years till Midsummer 1825, when I was sixteen years old. I boarded at this school, so that I had the great advantage of living the life of a true schoolboy ; but as the distance was hardly more than a mile to my home, I very often ran there in the longer intervals between the callings over and before locking up at night. This, I think, was in many ways advantageous to me by keeping up home affections and interests. I remember in the early part of my school life that I often had to run very quickly to be in time, and from being a fleet runner was generally successful ; but when in doubt I prayed earnestly to God to help me, and I well remember that I attributed my success to the prayers and not to my quick running, and marvelled how generally I was aided.

I have heard my father and elder sister say that I had, as a very young boy, a strong taste for long solitary walks ; but what I thought about I know not. I often became quite absorbed, and once, whilst returning to school on the summit of the old fortifications round Shrewsbury, which had been converted into a public foot-path with no parapet on one side, I walked off and fell to the ground, but the height was only seven or eight feet. Nevertheless the number of thoughts which passed through my mind during this very short, but sudden and wholly unexpected fall, was astonishing, and seem hardly compatible with what physiologists have, I believe, proved about each thought requiring quite an appreciable amount of time.

Nothing could have been worse for the development of my mind than Dr. Butler's school, as it was strictly classical, nothing else being taught, except a little ancient geography and history. The school as a means of education to me was simply a blank. During my whole life I have been singularly incapable of mastering any language. Especial attention was paid to verse-making, and this I could never do well. I had many friends, and got together a good collection of old verses, which by patching together, sometimes aided by other boys, I could work into any subject. Much attention was paid to learning by heart the lessons of the previous day; this I could effect with great facility, learning forty or fifty lines of Virgil or Homer, whilst I was in morning chapel; but this exercise was utterly useless, for every verse was forgotten in forty-eight hours. I was not idle, and with the exception of versification, generally worked conscientiously at my classics, not using cribs. The sole pleasure I ever received from such studies, was from some of the odes of Horace, which I admired greatly.

When I left the school I was for my age neither high nor low in it; and I believe that I was considered by all my masters and by my father as a very ordinary boy, rather below the common standard in intellect. To my deep mortification my father once said to me, "You care for nothing but shooting, dogs, and rat-catching, and you will be a disgrace to yourself and all your family." But my father, who was the kindest man I ever knew and whose memory I love with all my heart, must have been angry and somewhat unjust when he used such words.

Looking back as well as I can at my character during my school life, the only qualities which at this period promised well for the future, were, that I had strong and diversified tastes, much zeal for whatever interested me, and a keen pleasure in understanding any complex subject or thing. I was taught Euclid by a private tutor, and I distinctly remember the intense satisfaction which the clear geometrical proofs gave me. I remember with equal distinctness the delight which my uncle gave me (the father of Francis Galton) by explaining the principle of the vernier of a barometer. With respect to diversified tastes, independently of science, I was fond of reading various books, and I used to sit for hours reading the historical plays of Shakespeare, generally in an old window in the thick walls of the school. I read also other poetry, such as Thomson's 'Seasons,' and the recently published poems of Byron and Scott. I mention this because later in life I wholly lost, to my great regret, all pleasure from poetry of any kind, including Shakespeare. In connection with pleasure from poetry, I may add that in 1822 a vivid delight in scenery was first awakened in my mind, during a riding tour on the borders of Wales, and this has lasted longer than any other æsthetic pleasure.

Early in my school days a boy had a copy of the 'Wonders of the World,' which I often read, and disputed with other boys about the veracity of some of the statements; and I believe that this book first gave me a wish to travel in remote countries, which was ultimately fulfilled by the voyage of the *Beagle*. In the latter part of my school life I became passionately fond of shooting ; I do not believe that any one could have shown more zeal for the most holy cause than I did for shooting birds. How well I remember killing my first snipe, and my excitement was so great that I had much difficulty in reloading my gun from the trembling of my hands. This taste long continued, and I became a very good shot. When at Cambridge I used to practise throwing up my gun to my shoulder before a looking-glass to see that I threw it up straight. Another and better plan was to get a friend to wave about a lighted candle, and then to fire at it with a cap on the nipple, and if the aim was accurate the little puff of air would blow out the candle. The explosion of the cap caused a sharp crack, and I was told that the tutor of the college remarked, "What an extraordinary thing it is, Mr. Darwin seems to spend hours in cracking a horse-whip in his room, for I often hear the crack when I pass under his windows."

I had many friends amongst the schoolboys, whom I loved dearly, and I think that my disposition was then very affectionate.

With respect to science, I continued collecting minerals with much zeal, but quite unscientifically— all that I cared about was a new-*named* mineral, and I hardly attempted to classify them. I must have observed insects with some little care, for when ten years old (1819) I went for three weeks to Plas Edwards on the sea-coast in Wales, I was very much interested and surprised at seeing a large black and scarlet

Hemipterous insect, many moths (Zygæna), and a Cicindela which are not found in Shropshire. I almost made up my mind to begin collecting all the insects which I could find dead, for on consulting my sister I concluded that it was not right to kill insects for the sake of making a collection. From reading White's 'Selborne,' I took much pleasure in watching the habits of birds, and even made notes on the subject. In my simplicity I remember wondering why every gentleman did not become an ornithologist.

Towards the close of my school life, my brother worked hard at chemistry, and made a fair laboratory with proper apparatus in the tool-house in the garden, and I was allowed to aid him as a servant in most of his experiments. He made all the gases and many compounds, and I read with care several books on chemistry, such as Henry and Parkes' 'Chemical Catechism.' The subject interested me greatly, and we often used to go on working till rather late at night. This was the best part of my education at school, for it showed me practically the meaning of experimental science. The fact that we worked at chemistry somehow got known at school, and as it was an unprecedented fact, I was nicknamed "Gas." I was also once publicly rebuked by the head-master, Dr. Butler, for thus wasting my time on such useless subjects ; and he called me very unjustly a " poco curante," and as I did not understand what he meant, it seemed to me a fearful reproach.

As I was doing no good at school, my father wisely took me away at a rather earlier age than usual, and

D 2

sent me (Oct. 1825) to Edinburgh University with
my brother, where I stayed for two years or sessions.
My brother was completing his medical studies, though
I do not believe he ever really intended to practise,
and I was sent there to commence them. But soon
after this period I became convinced from various
small circumstances that my father would leave me
property enough to subsist on with some comfort,
though I never imagined that I should be so rich a
man as I am ; but my belief was sufficient to check
any strenuous effort to learn medicine.

The instruction at Edinburgh was altogether by
lectures, and these were intolerably dull, with the
exception of those on chemistry by Hope ; but to my
mind there are no advantages and many disadvantages
in lectures compared with reading. Dr. Duncan's
lectures on Materia Medica at 8 o'clock on a
winter's morning are something fearful to remember.
Dr. —— made his lectures on human anatomy as dull
as he was himself, and the subject disgusted me. It
has proved one of the greatest evils in my life that I
was not urged to practise dissection, for I should soon
have got over my disgust; and the practice would
have been invaluable for all my future work. This
has been an irremediable evil, as well as my inca-
pacity to draw. I also attended regularly the clinical
wards in the hospital. Some of the cases distressed
me a good deal, and I still have vivid pictures before
me of some of them ; but I was not so foolish as to
allow this to lessen my attendance. I cannot understand why this part of my medical course did not

interest me in a greater degree; for during the
summer before coming to Edinburgh I began attending some of the poor people, chiefly children and
women in Shrewsbury : I wrote down as full an
account as I could of the case with all the symptoms,
and read them aloud to my father, who suggested
further inquiries and advised me what medicines to
give, which I made up myself. At one time I had at
least a dozen patients, and I felt a keen interest in the
work. My father, who was by far the best judge of
character whom I ever knew, declared that I should
make a successful physician,—meaning by this one
who would get many patients. He maintained that
the chief element of success was exciting confidence;
but what he saw in me which convinced him that I
should create confidence I know not. I also attended
on two occasions the operating theatre in the hospital
at Edinburgh, and saw two very bad operations, one
on a child, but I rushed away before they were completed. Nor did I ever attend again, for hardly any
inducement would have been strong enough to make
me do so; this being long before the blessed days of
chloroform. The two cases fairly haunted me for
many a long year.

My brother stayed only one year at the University, so that during the second year I was left to
my own resources; and this was an advantage,
for I became well acquainted with several young
men fond of natural science. One of these was
Ainsworth, who afterwards published his travels in
Assyria ; he was a Wernerian geologist, and knew a

little about many subjects. Dr. Coldstream was a
very different young man, prim, formal, highly religious, and most kind-hearted; he afterwards published some good zoological articles. A third young
man was Hardie, who would, I think, have made a
good botanist, but died early in India. Lastly, Dr.
Grant, my senior by several years, but how I became
acquainted with him I cannot remember; he published
some first-rate zoological papers, but after coming to
London as Professor in University College, he did
nothing more in science, a fact which has always been
inexplicable to me. I knew him well; he was dry
and formal in manner, with much enthusiasm beneath
this outer crust. He one day, when we were walking
together, burst forth in high admiration of Lamarck
and his views on evolution. I listened in silent astonishment, and as far as I can judge without any
effect on my mind. I had previously read the ' Zoonomia' of my grandfather, in which similar views are
maintained, but without producing any effect on me.
Nevertheless it is probable that the hearing rather
early in life such views maintained and praised may
have favoured my upholding them under a different
form in my ' Origin of Species.' At this time I
admired greatly the ' Zoonomia;' but on reading it a
second time after an interval of ten or fifteen years, I
was much disappointed; the proportion of speculation
being so large to the facts given.

Drs. Grant and Coldstream attended much to
marine Zoology, and I often accompanied the former
to collect animals in the tidal pools, which I dissected

as well as I could. I also became friends with some of the Newhaven fishermen, and sometimes accompanied them when they trawled for oysters, and thus got many specimens. But from not having had any regular practice in dissection, and from possessing only a wretched miscroscope, my attempts were very poor. Nevertheless I made one interesting little discovery, and read, about the beginning of the year 1826, a short paper on the subject before the Plinian Society. This was that the so-called ova of Flustra had the power of independent movement by means of cilia, and were in fact larvæ. In another short paper I showed that the little globular bodies which had been supposed to be the young state of *Fucus loreus* were the egg-cases of the worm-like *Pontobdella muricata*.

The Plinian Society was encouraged and, I believe, founded by Professor Jameson : it consisted of students and met in an underground room in the University for the sake of reading papers on natural science and discussing them. I used regularly to attend, and the meetings had a good effect on me in stimulating my zeal and giving me new congenial acquaintances. One evening a poor young man got up, and after stammering for a prodigious length of time, blushing crimson, he at last slowly got out the words, "Mr. President, I have forgotten what I was going to say." The poor fellow looked quite overwhelmed, and all the members were so surprised that no one could think of a word to say to cover his confusion. The papers which were read to our little society were not printed, so that I had not the satis-

faction of seeing my paper in print; but I believe
Dr. Grant noticed my small discovery in his excellent·
memoir on Flustra.

I was also a member of the Royal Medical Society,
and attended pretty regularly; but as the subjects were
exclusively medical, I did not much care about them.
Much rubbish was talked there, but there were some
good speakers, of whom the best was the present Sir
J. Kay-Shuttleworth. Dr. Grant took me occasionally to the meetings of the Wernerian Society, where
various papers on natural history were read, discussed,
and afterwards published in the 'Transactions.' I
heard Audubon deliver there some interesting dis-
courses on the habits of N. American birds, sneering
somewhat unjustly at Waterton. By the way, a
negro lived in Edinburgh, who had travelled with
Waterton, and gained his livelihood by stuffing birds,
which he did excellently : he gave me lessons for
payment, and I used often to sit with him, for he was
a very pleasant and intelligent man.

Mr. Leonard Horner also took me once to a meeting
of the Royal Society of Edinburgh, where I saw Sir
Walter Scott in the chair as President, and he apolo-
gised to the meeting as not feeling fitted for such a
position. I looked at him and at the whole scene with
some awe and reverence, and I think it was owing to
this visit during my youth, and to my having attended
the Royal Medical Society, that I felt the honour of
being elected a few years ago an honorary member
of both these Societies, more than any other similar
honour. If I had been told at that time that I should

one day have been thus honoured, I declare that I should have thought it as ridiculous and improbable, as if I had been told that I should be elected King of England.

During my second year at Edinburgh I attended ——'s lectures on Geology and Zoology, but they were incredibly dull. The sole effect they produced on me was the determination never as long as I lived to read a book on Geology, or in any way to study the science. Yet I feel sure that I was prepared for a philosophical treatment of the subject; for an old Mr. Cotton in Shropshire, who knew a good deal about rocks, had pointed out to me two or three years previously a well-known large erratic boulder in the town of Shrewsbury, called the "bell-stone"; he told me that there was no rock of the same kind nearer than Cumberland or Scotland, and he solemnly assured me that the world would come to an end before any one would be able to explain how this stone came where it now lay. This produced a deep impression on me, and I meditated over this wonderful stone. So that I felt the keenest delight when I first read of the action of icebergs in transporting boulders, and I gloried in the progress of Geology. Equally striking is the fact that I, though now only sixty-seven years old, heard the Professor, in a field lecture at Salisbury Craigs, discoursing on a trap-dyke, with amygdaloidal margins and the strata indurated on each side, with volcanic rocks all around us, say that it was a fissure filled with sediment from above, adding with a sneer that there were men who main-

tained that it had been injected from beneath in a
molten condition. When I think of this lecture, I do
not wonder that I determined never to attend to
Geology.

From attending ——'s lectures, I became acquainted
with the curator of the museum, Mr. Macgillivray,
who afterwards published a large and excellent book
on the birds of Scotland. I had much interesting
natural-history talk with him, and he was very kind to
me. He gave me some rare shells, for I at that time
collected marine mollusca, but with no great zeal.

My summer vacations during these two years were
wholly given up to amusements, though I always had
some book in hand, which I read with interest.
During the summer of 1826 I took a long walking
tour with two friends with knapsacks on our backs
through North Wales. We walked thirty miles most
days, including one day the ascent of Snowdon. I
also went with my sister a riding tour in North Wales,
a servant with saddle-bags carrying our clothes. The
autumns were devoted to shooting chiefly at Mr.
Owen's, at Woodhouse, and at my Uncle Jos's, * at
Maer. My zeal was so great that I used to place my
shooting-boots open by my bed-side when I went to
bed, so as not to lose half a minute in putting them on
in the morning; and on one occasion I reached a
distant part of the Maer estate, on the 20th of August
for black-game shooting, before I could see : I then
toiled on with the gamekeeper the whole day through
thick heath and young Scotch firs.

* Josiah Wedgwood, the son of the founder of the Etruria Works.

I kept an exact record of every bird which I shot throughout the whole season. One day when shooting at Woodhouse with Captain Owen, the eldest son, and Major Hill, his cousin, afterwards Lord Berwick, both of whom I liked very much, I thought myself shamefully used, for every time after I had fired and thought that I had killed a bird, one of the two acted as if loading his gun, and cried out, "You must not count that bird, for I fired at the same time," and the gamekeeper, perceiving the joke, backed them up. After some hours they told me the joke, but it was no joke to me, for I had shot a large number of birds, but did not know how many, and could not add them to my list, which I used to do by making a knot in a piece of string tied to a button-hole. This my wicked friends had perceived.

How I did enjoy shooting! but I think that I must have been half-consciously ashamed of my zeal, for I tried to persuade myself that shooting was almost an intellectual employment; it required so much skill to judge where to find most game and to hunt the dogs well.

One of my autumnal visits to Maer in 1827 was memorable from meeting there Sir J. Mackintosh, who was the best converser I ever listened to. I heard afterwards with a glow of pride that he had said, "There is something in that young man that interests me." This must have been chiefly due to his perceiving that I listened with much interest to everything which he said, for I was as ignorant as a pig about his subjects of history, politics, and moral

philosophy. To hear of praise from an eminent
person, though no doubt apt or certain to excite
vanity, is, I think, good for a young man, as it helps
to keep him in the right course.

My visits to Maer during these two or three suc-
ceeding years were quite delightful, independently of
the autumnal shooting. Life there was perfectly
free; the country was very pleasant for walking or
riding; and in the evening there was much very
agreeable conversation, not so personal as it generally
is in large family parties, together with music. In the
summer the whole family used often to sit on the
steps of the old portico, with the flower-garden in
front, and with the steep wooded bank opposite the
house reflected in the lake, with here and there a fish
rising or a water-bird paddling about. Nothing has
left a more vivid picture on my mind than these
evenings at Maer. I was also attached to and greatly
revered my Uncle Jos; he was silent and reserved, so
as to be a rather awful man; but he sometimes talked
openly with me. He was the very type of an upright
man, with the clearest judgment. I do not believe
that any power on earth could have made him swerve
an inch from what he considered the right course. I
used to apply to him in my mind the well-known ode
of Horace, now forgotten by me, in which the words
" nec vultus tyranni, &c.," * come in.

Cambridge 1828–1831.—After having spent two

* Justum et tenacem propositi virum
Non civium ardor prava jubentium,
Non vultus instantis tyranni
Mente quatit solidâ.

sessions in Edinburgh, my father perceived, or he heard from my sisters, that I did not like the thought of being a physician, so he proposed that I should become a clergyman. He was very properly vehement against my turning into an idle sporting man, which then seemed my probable destination. I asked for some time to consider, as from what little I had heard or thought on the subject I had scruples about declaring my belief in all the dogmas of the Church of England ; though otherwise I liked the thought of being a country clergyman. Accordingly I read with care ' Pearson on the Creeds,' and a few other books on divinity ; and as I did not then in the least doubt the strict and literal truth of every word in the Bible, I soon persuaded myself that our Creed must be fully accepted.

Considering how fiercely I have been attacked by the orthodox, it seems ludicrous that I once intended to be a clergyman. Nor was this intention and my father's wish ever formally given up, but died a natural death when, on leaving Cambridge, I joined the *Beagle* as naturalist. If the phrenologists are to be trusted, I was well fitted in one respect to be a clergyman. A few years ago the secretaries of a German psychological society asked me earnestly by letter for a photograph of myself ; and some time afterwards I received the proceedings of one of the meetings, in which it seemed that the shape of my head had been the subject of a public discussion, and one of the speakers declared that I had the bump of reverence developed enough for ten priests.

As it was decided that I should be a clergyman, it was necessary that I should go to one of the English universities and take a degree; but as I had never opened a classical book since leaving school, I found to my dismay, that in the two intervening years I had actually forgotten, incredible as it may appear, almost everything which I had learnt, even to some few of the Greek letters. I did not therefore proceed to Cambridge at the usual time in October, but worked with a private tutor in Shrewsbury, and went to Cambridge after the Christmas vacation, early in 1828. I soon recovered my school standard of knowledge, and could translate easy Greek books, such as Homer and the Greek Testament, with moderate facility.

During the three years which I spent at Cambridge my time was wasted, as far as the academical studies were concerned, as completely as at Edinburgh and at school. I attempted mathematics, and even went during the summer of 1828 with a private tutor (a very dull man) to Barmouth, but I got on very slowly. The work was repugnant to me, chiefly from my not being able to see any meaning in the early steps in algebra. This impatience was very foolish, and in after years I have deeply regretted that I did not proceed far enough at least to understand something of the great leading principles of mathematics, for men thus endowed seem to have an extra sense. But I do not believe that I should ever have succeeded beyond a very low grade. With respect to Classics I did nothing except attend a few compulsory college

lectures, and the attendance was almost nominal. In
my second year I had to work for a month or two to
pass the Little-Go, which I did easily. Again, in my
last year I worked with some earnestness for my final
degree of B.A., and brushed up my Classics, together
with a little Algebra and Euclid, which latter gave me
much pleasure, as it did at school. In order to pass
the B.A. examination, it was also necessary to get up
Paley's 'Evidences of Christianity,' and his 'Moral
Philosophy.' This was done in a thorough manner,
and I am convinced that I could have written out the
whole of the 'Evidences' with perfect correctness, but
not of course in the clear language of Paley. The
logic of this book and, as I may add, of his 'Natural
Theology,' gave me as much delight as did Euclid.
The careful study of these works, without attempting
to learn any part by rote, was the only part of the
academical course which, as I then felt and as I still
believe, was of the least use to me in the education of
my mind. I did not at that time trouble myself about
Paley's premises; and taking these on trust, I was
charmed and convinced by the long line of argumen-
tation. By answering well the examination questions
in Paley, by doing Euclid well, and by not failing
miserably in Classics, I gained a good place among
the οἱ πολλοὶ or crowd of men who do not go in for
honours. Oddly enough, I cannot remember how
high I stood, and my memory fluctuates between the
fifth, tenth, or twelfth, name on the list.*

Public lectures on several branches were given in

* Tenth in the list of January 1831.

the University, attendance being quite voluntary ; but I was so sickened with lectures at Edinburgh that I did not even attend Sedgwick's eloquent and interesting lectures. Had I done so I should probably have become a geologist earlier than I did. I attended, however, Henslow's lectures on Botany, and liked them much for their extreme clearness, and the admirable illustrations ; but I did not study botany. Henslow used to take his pupils, including several of the older members of the University, field excursions, on foot or in coaches, to distant places, or in a barge down the river, and lectured on the rarer plants and animals which were observed. These excursions were delightful.

Although, as we shall presently see, there were some redeeming features in my life at Cambridge, my time was sadly wasted there, and worse than wasted. From my passion for shooting and for hunting, and, when this failed, for riding across country, I got into a sporting set, including some dissipated low-minded young men. We used often to dine together in the evening, though these dinners often included men of a higher stamp, and we sometimes drank too much, with jolly singing and playing at cards afterwards. I know that I ought to feel ashamed of days and evenings thus spent, but as some of my friends were very pleasant, and we were all in the highest spirits, I cannot help looking back to these times with much pleasure.

But I am glad to think that I had many other friends of a widely different nature. I was very intimate with Whitley,* who was afterwards Senior Wrangler, and we used continually to take long walks together. He inoculated me with a taste for pictures and good engravings, of which I bought some. I frequently went to the Fitzwilliam Gallery, and my taste must have been fairly good, for I certainly admired the best pictures, which I discussed with the old curator. I read also with much interest Sir Joshua Reynolds' book. This taste, though not natural to me, lasted for several years, and many of the pictures in the National Gallery in London gave me much pleasure ; that of Sebastian del Piombo exciting in me a sense of sublimity.

I also got into a musical set, I believe by means of my warm-hearted friend, Herbert,† who took a high wrangler's degree. From associating with these men, and hearing them play, I acquired a strong taste for music, and used very often to time my walks so as to hear on week days the anthem in King's College Chapel. This gave me intense pleasure, so that my backbone would sometimes shiver. I am sure that there was no affectation or mere imitation in this taste, for I used generally to go by myself to King's College, and I sometimes hired the chorister boys to sing in my rooms. Nevertheless I am so utterly destitute of an ear, that I cannot perceive a discord, or keep time and hum a tune correctly ; and it is a mystery how I could possibly have derived pleasure from music.

* Rev. C. Whitley, Hon. Canon of Durham, formerly Reader in Natural Philosophy in Durham University.

† The late John Maurice Herbert, County Court Judge of Cardiff.and the Monmouth Circuit.

My musical friends soon perceived my state, and sometimes amused themselves by making me pass an examination, which consisted in ascertaining how many tunes I could recognise, when they were played rather more quickly or slowly than usual. 'God save the King,' when thus played, was a sore puzzle. There was another man with almost as bad an ear as I had, and strange to say he played a little on the flute. Once I had the triumph of beating him in one of our musical examinations.

But no pursuit at Cambridge was followed with nearly so much eagerness or gave me so much pleasure as collecting beetles. It was the mere passion for collecting, for I did not dissect them, and rarely compared their external characters with published descriptions, but got them named anyhow. I will give a proof of my zeal: one day, on tearing off some old bark, I saw two rare beetles, and seized one in each hand; then I saw a third and new kind, which I could not bear to lose, so that I popped the one which I held in my right hand into my mouth. Alas! it ejected some intensely acrid fluid, which burnt my tongue so that I was forced to spit the beetle out, which was lost, as was the third one.

I was very successful in collecting, and invented two new methods; I employed a labourer to scrape during the winter, moss off old trees and place it in a large bag, and likewise to collect the rubbish at the bottom of the barges in which reeds are brought from the fens, and thus I got some very rare species. No poet ever felt more delighted at seeing his first poem published than I did at seeing, in Stephens' ' Illustrations of British Insects,' the magic words, "captured by C. Darwin, Esq." I was introduced to entomology by my second cousin, W. Darwin Fox, a clever and most pleasant man, who was then at Christ's College, and with whom I became extremely intimate. Afterwards I became well acquainted, and went out collecting, with Albert Way of Trinity, who in after years became a well-known archæologist; also with H. Thompson of the same College, afterwards a leading agriculturist, chairman of a great railway, and Member of Parliament. It seems therefore that a taste for collecting beetles is some indication of future success in life!

I am surprised what an indelible impression many of the beetles which I caught at Cambridge have left on my mind. I can remember the exact appearance of certain posts, old trees and banks where I made a good capture. The pretty *Panagæus crux-major* was a treasure in those days, and here at Down I saw a beetle running across a walk, and on picking it up instantly perceived that it differed slightly from *P. crux-major*, and it turned out to be *P. quadripunctatus*, which is only a variety or closely allied species, differing from it very slightly in outline. I had never seen in those old days Licinus alive, which to an uneducated eye hardly differs from many of the black Carabidous beetles; but my sons found here a specimen, and I instantly recognised that it was new to me; yet I had not looked at a British beetle for the last twenty years.

I have not as yet mentioned a circumstance which influenced my whole career more than any other. This was my friendship with Professor Henslow. Before coming up to Cambridge, I had heard of him from my brother as a man who knew every branch of science, and I was accordingly prepared to reverence him. He kept open house once every week when all undergraduates and some older members of the University, who were attached to science, used to meet in the evening. I soon got, through Fox, an invitation, and went there regularly. Before long I became well acquainted with Henslow, and during the latter half of my time at Cambridge took long walks with him on most days ; so that I was called by some of the dons " the man who walks with Henslow ; " and in the evening I was very often asked to join his family dinner. His knowledge was great in botany, entomology, chemistry, mineralogy, and geology. His strongest taste was to draw conclusions from long-continued minute observations. His judgment was excellent, and his whole mind well balanced ; but I do not suppose that any one would say that he possessed much original genius.

He was deeply religious, and so orthodox, that he told me one day he should be grieved if a single word of the Thirty-nine Articles were altered. His moral qualities were in every way admirable. He was free from every tinge of vanity or other petty feeling ; and I never saw a man who thought so little about himself or his own concerns. His temper was imperturbably good, with the most winning and courteous manners ; yet, as I have seen, he could be roused by any bad action to the warmest indignation and prompt action.

I once saw in his company in the streets of Cambridge almost as horrid a scene as could have been witnessed during the French Revolution. Two body-snatchers had been arrested, and whilst being taken to prison had been torn from the constable by a crowd of the roughest men, who dragged them by their legs along the muddy and stony road. They were covered from head to foot with mud, and their faces were bleeding either from having been kicked or from the stones ; they looked like corpses, but the crowd was so dense that I got only a few momentary glimpses of the wretched creatures. Never in my life have I seen such wrath painted on a man's face as was shown by Henslow at this horrid scene. He tried repeatedly to penetrate the mob ; but it was simply impossible. He then rushed away to the mayor, telling me not to follow him, but to get more policemen. I forget the issue, except that the two men were got into the prison without being killed.

Henslow's benevolence was unbounded, as he proved by his many excellent schemes for his poor parishioners, when in after years he held the living of Hitcham. My intimacy with such a man ought to have been, and I hope was, an inestimable benefit. I cannot resist mentioning a trifling incident, which showed his kind consideration. Whilst examining some pollen-grains on a damp surface, I saw the tubes exserted, and instantly rushed off to communicate my

surprising discovery to him. Now I do not suppose
any other professor of botany could have helped
laughing at my coming in such a hurry to make such
a communication. But he agreed how interesting the
phenomenon was, and explained its meaning, but
made me clearly understand how well it was known ;
so I left him not in the least mortified, but well
pleased at having discovered for myself so remarkable
a fact, but determined not to be in such a hurry again
to communicate my discoveries.

Dr. Whewell was one of the older and distinguished
men who sometimes visited Henslow, and on several
occasions I walked home with him at night. Next to
Sir J. Mackintosh he was the best converser on grave
subjects to whom I ever listened. Leonard Jenyns,*
who afterwards published some good essays in Natural
History,† often stayed with Henslow, who was his
brother-in-law. I visited him at his parsonage on
the borders of the Fens [Swaffham Bulbeck], and
had many a good walk and talk with him about
Natural History. I became also acquainted with
several other men older than me, who did not care
much about science, but were friends of Henslow.
One was a Scotchman, brother of Sir Alexander
Ramsay, and tutor of Jesus College ; he was a delightful man, but did not live for many years. Another
was Mr. Dawes, afterwards Dean of Hereford, and
famous for his success in the education of the poor.

* The well-known Soame Jenyns
was cousin to Mr. Jenyns' father.
† Mr. Jenyns (now Blomefield)
described the fish for the Zoology
of the *Beagle;* and is author of a
long series of papers, chiefly Zoological.

These men and others of the same standing, together with Henslow, used sometimes to take distant excursions into the country, which I was allowed to join, and they were most agreeable.

Looking back, I infer that there must have been something in me a little superior to the common run of youths, otherwise the above-mentioned men, so much older than me and higher in academical position, would never have allowed me to associate with them. Certainly I was not aware of any such superiority, and I remember one of my sporting friends, Turner, who saw me at work with my beetles, saying that I should some day be a Fellow of the Royal Society, and the notion seemed to me preposterous.

During my last year at Cambridge, I read with care and profound interest Humboldt's 'Personal Narrative.' This work, and Sir J. Herschel's 'Introduction to the Study of Natural Philosophy,' stirred up in me a burning zeal to add even the most humble contribution to the noble structure of Natural Science. No one or a dozen other books influenced me nearly so much as these two. I copied out from Humboldt long passages about Teneriffe, and read them aloud on one of the above-mentioned excursions, to (I think) Henslow, Ramsay, and Dawes, for on a previous occasion I had talked about the glories of Teneriffe, and some of the party declared they would endeavour to go there; but I think that they were only half in earnest. I was, however, quite in earnest, and got an introduction to a merchant in London to enquire about ships; but the scheme was, of course, knocked on the head by the voyage of the *Beagle.*

My summer vacations were given up to collecting beetles, to some reading, and short tours. In the autumn my whole time was devoted to shooting, chiefly at Woodhouse and Maer, and sometimes with young Eyton of Eyton. Upon the whole the three years which I spent at Cambridge were the most joyful in my happy life; for I was then in excellent health, and almost always in high spirits.

As I had at first come up to Cambridge at Christmas, I was forced to keep two terms after passing my final examination, at the commencement of 1831; and Henslow then persuaded me to begin the study of geology. Therefore on my return to Shropshire I examined sections, and coloured a map of parts round Shrewsbury. Professor Sedgwick intended to visit North Wales in the beginning of August to pursue his famous geological investigations amongst the older rocks, and Henslow asked him to allow me to accompany him.* Accordingly he came and slept at my father's house.

A short conversation with him during this evening produced a strong impression on my mind. Whilst examining an old gravel-pit near Shrewsbury, a

---

* In connection with this tour my father used to tell a story about Sedgwick: they had started from their inn one morning, and had walked a mile or two, when Sedgwick suddenly stopped, and vowed that he would return, being certain "that damned scoundrel" (the waiter) had not given the chambermaid the sixpence intrusted to him for the purpose. He was ultimately persuaded to give up the project, seeing that there was no reason for suspecting the waiter of especial perfidy.—F. D.

labourer told me that he had found in it a large worn
tropical Volute shell, such as may be seen on the
chimney-pieces of cottages ; and as he would not sell
the shell, I was convinced that he had really found it
in the pit. I told Sedgwick of the fact, and he at
once said (no doubt truly) that it must have been
thrown away by some one into the pit ; but then
added, if really embedded there it would be the
greatest misfortune to geology, as it would overthrow
all that we know about the superficial deposits of
the Midland Counties. These gravel-beds belong in
fact to the glacial period, and in after years I found in
them broken arctic shells. But I was then utterly
astonished at Sedgwick not being delighted at so
wonderful a fact as a tropical shell being found near
the surface in the middle of England. Nothing
before had ever made me thoroughly realise, though
I had read various scientific books, that science consists in grouping facts so that general laws or conclusions may be drawn from them.

Next morning we started for Llangollen, Conway,
Bangor, and Capel Curig. This tour was of decided
use in teaching me a little how to make out the
geology of a country. Sedgwick often sent me on
a line parallel to his, telling me to bring back specimens of the rocks and to mark the stratification on a
map. I have little doubt that he did this for my
good, as I was too ignorant to have aided him. On
this tour I had a striking instance how easy it is to
overlook phenomena, however conspicuous, before
they have been observed by any one. We spent

many hours in Cwm Idwal, examining all the rocks
with extreme care, as Sedgwick was anxious to find
fossils in them ; but neither of us saw a trace of the
wonderful glacial phenomena all around us ; we did not
notice the plainly scored rocks, the perched boulders,
the lateral and terminal moraines.   Yet these phenomena are so conspicuous that, as I declared in a
paper published many years afterwards in the ' Philosophical Magazine,' * a house burnt down by fire did
not tell its story more plainly than did this valley.   If
it had still been filled by a glacier, the phenomena
would have been less distinct than they now are.

At Capel Curig I left Sedgwick and went in a
straight line by compass and map across the mountains to Barmouth, never following any track unless it
coincided with my course.   I thus came on some
strange wild places, and enjoyed much this manner of
travelling.   I visited Barmouth to see some Cambridge friends who were reading there, and thence
returned to Shrewsbury and to Maer for shooting ;
for at that time I should have thought myself mad to
give up the first days of partridge-shooting for geology
or any other science.

Voyage of the ' Beagle' from December 27, 1831, *to*
October 2, 1836.

On returning home from my short geological tour
in North Wales, I found a letter from Henslow, informing me that Captain Fitz-Roy was willing to give

* ' Philosophical Magazine,' 1842.

up part of his own cabin to any young man who would
volunteer to go with him without pay as naturalist to
the Voyage of the *Beagle*. I have given, as I believe,
in my MS. Journal an account of all the circumstances
which then occurred; I will here only say that I was
instantly eager to accept the offer, but my father
strongly objected, adding the words, fortunate for me,
" If you can find any man of common sense who
advises you to go I will give my consent." So I
wrote that evening and refused the offer. On the
next morning I went to Maer to be ready for September 1st, and, whilst out shooting, my uncle * sent
for me, offering to drive me over to Shrewsbury and
talk with my father, as my uncle thought it would be
wise in me to accept the offer. My father always
maintained that he was one of the most sensible men
in the world, and he at once consented in the kindest
manner. I had been rather extravagant at Cambridge, and to console my father, said, " that I
should be deuced clever to spend more than my
allowance whilst on board the *Beagle*;" but he answered with a smile, " But they tell me you are very
clever."

Next day I started for Cambridge to see Henslow,
and thence to London to see Fitz-Roy, and all was
soon arranged. Afterwards, on becoming very intimate with Fitz-Roy, I heard that I had run a very
narrow risk of being rejected, on account of the shape
of my nose! He was an ardent disciple of Lavater,
and was convinced that he could judge of a man's

* Josiah Wedgwood.

character by the outline of his features; and he
doubted whether any one with my nose could possess
sufficient energy and determination for the voyage.
But I think he was afterwards well satisfied that my
nose had spoken falsely.

Fitz-Roy's character was a singular one, with very
many noble features : he was devoted to his duty,
generous to a fault, bold, determined, and indomi-
tably energetic, and an ardent friend to all under his
sway. He would undertake any sort of trouble to
assist those whom he thought deserved assistance.
He was a handsome man, strikingly like a gentleman,
with highly courteous manners, which resembled those
of his maternal uncle, the famous Lord Castlereagh,
as I was told by the Minister at Rio. Nevertheless
he must have inherited much in his appearance from
Charles II., for Dr. Wallich gave me a collection of
photographs which he had made, and I was struck
with the resemblance of one to Fitz-Roy; and on
looking at the name, I found it Ch. E. Sobieski
Stuart, Count d'Albanie, a descendant of the same
monarch.

Fitz-Roy's temper was a most unfortunate one. It
was usually worst in the early morning, and with his
eagle eye he could generally detect something amiss
about the ship, and was then unsparing in his blame.
He was very kind to me, but was a man very difficult
to live with on the intimate terms which necessarily
followed from our messing by ourselves in the same
cabin. We had several quarrels; for instance, early in
the voyage at Bahia, in Brazil, he defended and praised

slavery, which I abominated, and told me that he had just visited a great slave-owner, who had called up many of his slaves and asked them whether they were happy, and whether they wished to be free, and all answered " No." I then asked him, perhaps with a sneer, whether he thought that the answer of slaves in the presence of their master was worth anything ? This made him excessively angry, and he said that as I doubted his word we could not live any longer together. I thought that I should have been compelled to leave the ship ; but as soon as the news spread, which it did quickly, as the captain sent for the first lieutenant to assuage his anger by abusing me, I was deeply gratified by receiving an invitation from all the gun-room officers to mess with them. But after a few hours Fitz-Roy showed his usual magnanimity by sending an officer to me with an apology and a request that I would continue to live with him.

His character was in several respects one of the most noble which I have ever known.

The voyage of the *Beagle* has been by far the most important event in my life, and has determined my whole career ; yet it depended on so small a circumstance as my uncle offering to drive me thirty miles to Shrewsbury, which few uncles would have done, and on such a trifle as the shape of my nose. I have always felt that I owe to the voyage the first real training or education of my mind ; I was led to attend closely to several branches of natural history, and thus my powers of observation

were improved, though they were always fairly
developed.

The investigation of the geology of all the places
visited was far more important, as reasoning here
comes into play. On first examining a new district
nothing can appear more hopeless than the chaos of
rocks ; but by recording the stratification and nature
of the rocks and fossils at many points, always reasoning and predicting what will be found elsewhere, light
soon begins to dawn on the district, and the structure
of the whole becomes more or less intelligible. I had
brought with me the first volume of Lyell's ' Principles
of Geology,' which I studied attentively; and the
book was of the highest service to me in many ways.
The very first place which I examined, namely
St. Jago in the Cape de Verde islands, showed me
clearly the wonderful superiority of Lyell's manner
of treating geology, compared with that of any other
author, whose works I had with me or ever afterwards
read.

Another of my occupations was collecting animals
of all classes, briefly describing and roughly dissecting
many of the marine ones ; but from not being able to
draw, and from not having sufficient anatomical knowledge, a great pile of MS. which I made during the
voyage has proved almost useless. I thus lost much
time, with the exception of that spent in acquiring
some knowledge of the Crustaceans, as this was of
service when in after years I undertook a monograph
of the Cirripedia.

During some part of the day I wrote my Journal,

and took much pains in describing carefully and vividly all that I had seen ; and this was good practice. My Journal served also, in part, as letters to my home, and portions were sent to England whenever there was an opportunity.

The above various special studies were, however, of no importance compared with the habit of energetic industry and of concentrated attention to whatever I was engaged in, which I then acquired. Everything about which I thought or read was made to bear directly on what I had seen or was likely to see ; and this habit of mind was continued during the five years of the voyage. I feel sure that it was this training which has enabled me to do whatever I have done in science.

Looking backwards, I can now perceive how my love for science gradually preponderated over every other taste. During the first two years my old passion for shooting survived in nearly full force, and I shot myself all the birds and animals for my collection ; but gradually I gave up my gun more and more, and finally altogether, to my servant, as shooting interfered with my work, more especially with making out the geological structure of a country. I discovered, though unconsciously and insensibly, that the pleasure of observing and reasoning was a much higher one than that of skill and sport. That my mind became developed through my pursuits during the voyage is rendered probable by a remark made by my father, who was the most acute observer whom I ever saw, of a sceptical disposition, and far from being a believer

in phrenology ; for on first seeing me after the voyage,
he turned round to my sisters, and exclaimed, "Why,
the shape of his head is quite altered."

To return to the voyage. On September 11th
(1831), I paid a flying visit with Fitz-Roy to the
Beagle at Plymouth. Thence to Shrewsbury to wish
my father and sisters a long farewell. On October
24th I took up my residence at Plymouth, and remained there until December 27th, when the *Beagle*
finally left the shores of England for her circumnavigation of the world. We made two earlier attempts
to sail, but were driven back each time by heavy
gales. These two months at Plymouth were the
most miserable which I ever spent, though I exerted
myself in various ways. I was out of spirits at the
thought of leaving all my family and friends for so
long a time, and the weather seemed to me inexpressibly gloomy. I was also troubled with palpitation
and pain about the heart, and like many a young
ignorant man, especially one with a smattering of
medical knowledge, was convinced that I had heart
disease. I did not consult any doctor, as I fully expected to hear the verdict that I was not fit for the
voyage, and I was resolved to go at all hazards.

I need not here refer to the events of the voyage—
where we went and what we did—as I have given a
sufficiently full account in my published Journal. The
glories of the vegetation of the Tropics rise before my
mind at the present time more vividly than anything
else ; though the sense of sublimity, which the great
deserts of Patagonia and the forest-clad mountains of

Tierra del Fuego excited in me, has left an indelible impression on my mind. The sight of a naked savage in his native land is an event which can never be forgotten. Many of my excursions on horseback through wild countries, or in the boats, some of which lasted several weeks, were deeply interesting; their discomfort and some degree of danger were at that time hardly a drawback, and none at all afterwards. I also reflect with high satisfaction on some of my scientific work, such as solving the problem of coral islands, and making out the geological structure of certain islands, for instance, St. Helena. Nor must I pass over the discovery of the singular relations of the animals and plants inhabiting the several islands of the Galapagos archipelago, and of all of them to the inhabitants of South America.

As far as I can judge of myself, I worked to the utmost during the voyage from the mere pleasure of investigation, and from my strong desire to add a few facts to the great mass of facts in Natural Science. But I was also ambitious to take a fair place among scientific men,—whether more ambitious or less so than most of my fellow-workers, I can form no opinion.

The geology of St. Jago is very striking, yet simple : a stream of lava formerly flowed over the bed of the sea, formed of triturated recent shells and corals, which it has baked into a hard white rock. Since then the whole island has been upheaved. But the line of white rock revealed to me a new and important fact, namely, that there had been afterwards subsi-

dence round the craters, which had since been in
action, and had poured forth lava. It then first
dawned on me that I might perhaps write a book on
the geology of the various countries visited, and this
made me thrill with delight. That was a memorable
hour to me, and how distinctly I can call to mind the
low cliff of lava beneath which I rested, with the sun
glaring hot, a few strange desert plants growing near,
and with living corals in the tidal pools at my feet.
Later in the voyage, Fitz-Roy asked me to read some
of my Journal, and declared it would be worth publishing ; so here was a second book in prospect !

Towards the close of our voyage I received a letter
whilst at Ascension, in which my sisters told me that
Sedgwick had called on my father, and said that I
should take a place among the leading scientific men.
I could not at the time understand how he could have
learnt anything of my proceedings, but I heard (I
believe afterwards) that Henslow had read some of
the letters which I wrote to him before the Philosophical Society of Cambridge,* and had printed them
for private distribution. My collection of fossil bones,
which had been sent to Henslow, also excited considerable attention amongst palæontologists. After
reading this letter, I clambered over the mountains of
Ascension with a bounding step, and made the volcanic
rocks resound under my geological hammer. All this
shows how ambitious I was ; but I think that I can

* Read at the meeting held
November 16, 1835, and printed in
a pamphlet of 31 pp. for distribution among the members of the
Society.

say with truth that in after years, though I cared in the highest degree for the approbation of such men as Lyell and Hooker, who were my friends, I did not care much about the general public. I do not mean to say that a favourable review or a large sale of my books did not please me greatly, but the pleasure was a fleeting one, and I am sure that I have never turned one inch out of my course to gain fame.

From my return to England (October 2, 1836) to my marriage (January 29, 1839).

These two years and three months were the most active ones which I ever spent, though I was occasionally unwell, and so lost some time. After going backwards and forwards several times between Shrewsbury, Maer, Cambridge, and London, I settled in lodgings at Cambridge * on December 13th, where all my collections were under the care of Henslow. I stayed here three months, and got my minerals and rocks examined by the aid of Professor Miller.

I began preparing my 'Journal of Travels,' which was not hard work, as my MS. Journal had been written with care, and my chief labour was making an abstract of my more interesting scientific results. I sent also, at the request of Lyell, a short account of my observations on the elevation of the coast of Chile to the Geological Society.†

On March 7th, 1837, I took lodgings in Great Marlborough Street in London, and remained there for

---

* In Fitzwilliam Street.

† 'Geolog. Soc. Proc.' ii. 1838, pp. 446-449.

nearly two years, until I was married. During these
two years I finished my Journal, read several papers
before the Geological Society, began preparing the
MS. for my 'Geological Observations,' and arranged
for the publication of the 'Zoology of the Voyage of
the *Beagle.*' In July I opened my first note-book for
facts in relation to the Origin of Species, about which
I had long reflected, and never ceased working for the
next twenty years.

During these two years I also went a little into
society, and acted as one of the honorary secretaries
of the Geological Society. I saw a great deal of Lyell.
One of his chief characteristics was his sympathy
with the work of others, and I was as much astonished as delighted at the interest which he showed
when, on my return to England, I explained to him
my views on coral reefs. This encouraged me greatly,
and his advice and example had much influence on me.
During this time I saw also a good deal of Robert
Brown ; I used often to call and sit with him during
his breakfast on Sunday mornings, and he poured
forth a rich treasure of curious observations and acute
remarks, but they almost always related to minute
points, and he never with me discussed large or
general questions in science.

During these two years I took several short excursions as a relaxation, and one longer one to the
Parallel Roads of Glen Roy, an account of which was
published in the 'Philosophical Transactions.'* This
paper was a great failure, and I am ashamed of it.

* 1839, pp. 39–82.

Having been deeply impressed with what I had seen of the elevation of the land in South America, I attributed the parallel lines to the action of the sea ; but I had to give up this view when Agassiz propounded his glacier-lake theory. Because no other explanation was possible under our then state of knowledge, I argued in favour of sea-action ; and my error has been a good lesson to me never to trust in science to the principle of exclusion.

As I was not able to work all day at science, I read a good deal during these two years on various subjects, including some metaphysical books; but I was not well fitted for such studies. About this time I took much delight in Wordsworth's and Coleridge's poetry ; and can boast that I read the 'Excursion' twice through. Formerly Milton's 'Paradise Lost' had been my chief favourite, and in my excursions during the voyage of the *Beagle*, when I could take only a single volume, I always chose Milton.

From my marriage, January 29, 1839, and residence in Upper Gower Street, to our leaving London and settling at Down, September 14, 1842.

After speaking of his happy married life, and of his children, he continues :—

During the three years and eight months whilst we resided in London, I did less scientific work, though I worked as hard as I possibly could, than during any other equal length of time in my life. This was owing to frequently recurring unwellness, and to one long and serious illness. The greater part of my

time, when I could do anything, was devoted to my
work on 'Coral Reefs,' which I had begun before
my marriage, and of which the last proof-sheet was
corrected on May 6th, 1842. This book, though a
small one, cost me twenty months of hard work, as
I had to read every work on the islands of the Pacific
and to consult many charts. It was thought highly of
by scientific men, and the theory therein given is,
I think, now well established.

No other work of mine was begun in so deductive
a spirit as this, for the whole theory was thought out
on the west coast of South America, before I had seen
a true coral reef. I had therefore only to verify and
extend my views by a careful examination of living
reefs. But it should be observed that I had during
the two previous years been incessantly attending to
the effects on the shores of South America of the
intermittent elevation of the land, together with
denudation and the deposition of sediment. This
necessarily led me to reflect much on the effects of
subsidence, and it was easy to replace in imagination
the continued deposition of sediment by the upward
growth of corals. To do this was to form my theory
of the formation of barrier-reefs and atolls.

Besides my work on coral-reefs, during my residence
in London, I read before the Geological Society
papers on the Erratic Boulders of South America,* on
Earthquakes,† and on the Formation by the Agency of
Earth-worms of Mould.‡ I also continued to superin-

* 'Geolog. Soc. Proc.' iii. 1842.   † 'Geolog. Trans.' v. 1840.
          ‡ 'Geolog. Soc. Proc.' ii. 1838.

tend the publication of the 'Zoology of the Voyage of the *Beagle.*' Nor did I ever intermit collecting facts bearing on the origin of species ; and I could sometimes do this when I could do nothing else from illness.

In the summer of 1842 I was stronger than I had been for some time, and took a little tour by myself in North Wales, for the sake of observing the effects of the old glaciers which formerly filled all the larger valleys. I published a short account of what I saw in the 'Philosophical Magazine.'* This excursion interested me greatly, and it was the last time I was ever strong enough to climb mountains or to take long walks such as are necessary for geological work.

During the early part of our life in London, I was strong enough to go into general society, and saw a good deal of several scientific men, and other more or less distinguished men. I will give my impressions with respect to some of them, though I have little to say worth saying.

I saw more of Lyell than of any other man, both before and after my marriage. His mind was characterised, as it appeared to me, by clearness, caution, sound judgment, and a good deal of originality. When I made any remark to him on Geology, he never rested until he saw the whole case clearly, and often made me see it more clearly than I had done before. He would advance all possible objections to my suggestion, and even after these were exhausted would long remain dubious. A second characteristic

* 'Philosophical Magazine,' 1842.

was his hearty sympathy with the work of other scientific men.*

On my return from the voyage of the *Beagle*, I explained to him my views on coral-reefs, which differed from his, and I was greatly surprised and encouraged by the vivid interest which he showed. His delight in science was ardent, and he felt the keenest interest in the future progress of mankind. He was very kind-hearted, and thoroughly liberal in his religious beliefs, or rather disbeliefs ; but he was a strong theist. His candour was highly remarkable. He exhibited this by becoming a convert to the Descent theory, though he had gained much fame by opposing Lamarck's views, and this after he had grown old. He reminded me that I had many years before said to him, when discussing the opposition of the old school of geologists to his new views, "What a good thing it would be if every scientific man was to die when sixty years old, as afterwards he would be sure to oppose all new doctrines." But he hoped that now he might be allowed to live.

The science of Geology is enormously indebted to Lyell—more so, as I believe, than to any other man who ever lived. When [I was] starting on the voyage of the *Beagle*, the sagacious Henslow, who, like all other geologists, believed at that time in successive cataclysms, advised me to get and study the first volume of the 'Principles,' which had then just been published,

---

* The slight repetition here observable is accounted for by the notes on Lyell, &c., having been added in April, 1881, a few years after the rest of the 'Recollections' were written.

but on no account to accept the views therein advocated. How differently would any one now speak of the 'Principles'! I am proud to remember that the first place, namely, St. Jago, in the Cape de Verde archipelago, in which I geologised, convinced me of the infinite superiority of Lyell's views over those advocated in any other work known to me.

The powerful effects of Lyell's works could formerly be plainly seen in the different progress of the science in France and England. The present total oblivion of Elie de Beaumont's wild hypotheses, such as his 'Craters of Elevation' and 'Lines of Elevation' (which latter hypothesis I heard Sedgwick at the Geological Society lauding to the skies), may be largely attributed to Lyell.

I saw a good deal of Robert Brown, "facile Princeps Botanicorum," as he was called by Humboldt. He seemed to me to be chiefly remarkable for the minuteness of his observations, and their perfect accuracy. His knowledge was extraordinarily great, and much died with him, owing to his excessive fear of ever making a mistake. He poured out his knowledge to me in the most unreserved manner, yet was strangely jealous on some points. I called on him two or three times before the voyage of the *Beagle*, and on one occasion he asked me to look through a microscope and describe what I saw. This I did, and believe now that it was the marvellous currents of protoplasm in some vegetable cell. I then asked him what I had seen ; but he answered me, " That is my little secret."

He was capable of the most generous actions. When old, much out of health, and quite unfit for any exertion, he daily visited (as Hooker told me) an old man-servant, who lived at a distance (and whom he supported), and read aloud to him. This is enough to make up for any degree of scientific penuriousness or jealousy.

I may here mention a few other eminent men, whom I have occasionally seen, but I have little to say about them worth saying. I felt a high reverence for Sir J. Herschel, and was delighted to dine with him at his charming house at the Cape of Good Hope, and afterwards at his London house. I saw him, also, on a few other occasions. He never talked much, but every word which he uttered was worth listening to.

I once met at breakfast at Sir R. Murchison's house the illustrious Humboldt, who honoured me by expressing a wish to see me. I was a little disappointed with the great man, but my anticipations probably were too high. I can remember nothing distinctly about our interview, except that Humboldt was very cheerful and talked much.

—— reminds me of Buckle whom I once met at Hensleigh Wedgwood's. I was very glad to learn from him his system of collecting facts. He told me that he bought all the books which he read, and made a full index, to each, of the facts which he thought might prove serviceable to him, and that he could always remember in what book he had read anything, for his memory was wonderful. I asked him how at first he could judge what facts would be serviceable, and he answered that he did not know, but that a sort of instinct guided him. From this habit of making indices, he was enabled to give the astonishing number of references on all sorts of subjects, which may be found in his 'History of Civilisation.' This book I thought most interesting, and read it twice, but I doubt whether his generalisations are worth anything. Buckle was a great talker, and I listened to him saying hardly a word, nor indeed could I have done so for he left no gaps. When Mrs. Farrer began to sing, I jumped up and said that I must listen to her; after I had moved away he turned round to a friend and said (as was overheard by my brother), "Well, Mr. Darwin's books are much better than his conversation."

Of other great literary men, I once met Sydney Smith at Dean Milman's house. There was something inexplicably amusing in every word which he uttered. Perhaps this was partly due to the expectation of being amused. He was talking about Lady Cork, who was then extremely old. This was the lady who, as he said, was once so much affected by one of his charity sermons, that she *borrowed* a guinea from a friend to put in the plate. He now said "It is generally believed that my dear old friend Lady Cork has been overlooked," and he said this in such a manner that no one could for a moment doubt that he meant that his dear old friend had been overlooked by the devil. How he managed to express this I know not.

I likewise once met Macaulay at Lord Stanhope's

(the historian's) house, and as there was only one
other man at dinner, I had a grand opportunity of
hearing him converse, and he was very agreeable.
He did not talk at all too much ; nor indeed could
such a man talk too much, as long as he allowed
others to turn the stream of his conversation, and this
he did allow.

Lord Stanhope once gave me a curious little proof
of the accuracy and fulness of Macaulay's memory :
many historians used often to meet at Lord Stanhope's
house, and in discussing various subjects they would
sometimes differ from Macaulay, and formerly they
often referred to some book to see who was right ;
but latterly, as Lord Stanhope noticed, no historian
ever took this trouble, and whatever Macaulay said
was final.

On another occasion I met at Lord Stanhope's
house, one of his parties of historians and other
literary men, and amongst them were Motley and
Grote. After luncheon I walked about Chevening
Park for nearly an hour with Grote, and was much
interested by his conversation and pleased by the
simplicity and absence of all pretension in his manners.

Long ago I dined occasionally with the old Earl,
the father of the historian ; he was a strange man,
but what little I knew of him I liked much. He was
frank, genial, and pleasant. He had strongly marked
features, with a brown complexion, and his clothes,
when I saw him, were all brown. He seemed to
believe in everything which was to others utterly incredible. He said one day to me, " Why don't you

give up your fiddle-faddle of geology and zoology, and turn to the occult sciences?" The historian, then Lord Mahon, seemed shocked at such a speech to me, and his charming wife much amused.

The last man whom I will mention is Carlyle, seen by me several times at my brother's house, and two or three times at my own house. His talk was very racy and interesting, just like his writings, but he sometimes went on too long on the same subject. I remember a funny dinner at my brother's, where, amongst a few others, were Babbage and Lyell, both of whom liked to talk. Carlyle, however, silenced every one by haranguing during the whole dinner on the advantages of silence. After dinner Babbage, in his grimmest manner, thanked Carlyle for his very interesting lecture on silence.

Carlyle sneered at almost every one: one day in my house he called Grote's 'History' "a fetid quagmire, with nothing spiritual about it." I always thought, until his 'Reminiscences' appeared, that his sneers were partly jokes, but this now seems rather doubtful. His expression was that of a depressed, almost despondent yet benevolent, man; and it is notorious how heartily he laughed. I believe that his benevolence was real, though stained by not a little jealousy. No one can doubt about his extraordinary power of drawing pictures of things and men—far more vivid, as it appears to me, than any drawn by Macaulay. Whether his pictures of men were true ones is another question.

He has been all-powerful in impressing some grand

moral truths on the minds of men. On the other
hand, his views about slavery were revolting. In his
eyes might was right. His mind seemed to me a very
narrow one ; even if all branches of science, which he
despised, are excluded. It is astonishing to me that
Kingsley should have spoken of him as a man well
fitted to advance science. He laughed to scorn the idea
that a mathematician, such as Whewell, could judge,
as I maintained he could, of Goethe's views on light.
He thought it a most ridiculous thing that any one
should care whether a glacier moved a little quicker
or a little slower, or moved at all. As far as I could
judge, I never met a man with a mind so ill adapted
for scientific research.

Whilst living in London, I attended as regularly as
I could the meetings of several scientific societies, and
acted as secretary to the Geological Society. But such
attendance, and ordinary society, suited my health
so badly that we resolved to live in the country, which
we both preferred and have never repented of.

Residence at Down from September 14, 1842, *to the
present time,* 1876.

After several fruitless searches in Surrey and
elsewhere, we found this house and purchased it.
I was pleased with the diversified appearance of
the vegetation proper to a chalk district, and so
unlike what I had been accustomed to in the Midland counties ; and still more pleased with the extreme quietness and rusticity of the place. It is

not, however, quite so retired a place as a writer in a German periodical makes it, who says that my house can be approached only by a mule-track ! Our fixing ourselves here has answered admirably in one way, which we did not anticipate, namely, by being very convenient for frequent visits from our children.

Few persons can have lived a more retired life than we have done. Besides short visits to the houses of relations, and occasionally to the seaside or elsewhere, we have gone nowhere. During the first part of our residence we went a little into society, and received a few friends here ; but my health almost always suffered from the excitement, violent shivering and vomiting attacks being thus brought on. I have therefore been compelled for many years to give up all dinner-parties ; and this has been somewhat of a deprivation to me, as such parties always put me into high spirits. From the same cause I have been able to invite here very few scientific acquaintances.

My chief enjoyment and sole employment throughout life has been scientific work ; and the excitement from such work makes me for the time forget, or drives quite away, my daily discomfort. I have therefore nothing to record during the rest of my life, except the publication of my several books. Perhaps a few details how they arose may be worth giving.

My several Publications.—In the early part of 1844, my observations on the volcanic islands visited during the voyage of the *Beagle* were published. In 1845, I took much pains in correcting a new edition of my ' Journal of Researches,' which was originally published in 1839 as part of Fitz-Roy's work. The success of this my first literary child always tickles my vanity more than that of any of my other books. , Even to this day it sells steadily in England and the United States, and has been translated for the second time into German, and into French and other languages. This success of a book of travels, especially of a scientific one, so many years after its first publication, is surprising. Ten thousand copies have been sold in England of the second edition. In 1846 my 'Geological Observations on South America' were published. I record in a little diary, which I have always kept, that my three geological books ('Coral Reefs' included) consumed four and a half years' steady work; "and now it is ten years since my return to England. How much time have I lost by illness?" I have nothing to say about these three books except that to my surprise new editions have lately been called for.*

In October, 1846, I began to work on 'Cirripedia.' When on the coast of Chile, I found a most curious form, which burrowed into the shells of Concholepas, and which differed so much from all other Cirripedes that I had to form a new sub-order for its sole reception. Lately an allied burrowing genus has been found on the shores of Portugal. To understand the structure of my new Cirripede I had to examine and dissect many of the common forms; and this gradually led me on to take up the whole group. I worked steadily on this subject for the next eight years, and ultimately

* 'Geological Observations,' 2nd Edit. 1876. 'Coral Reefs,' 2nd Edit. 1874.

published two thick volumes,* describing all the known living species, and two thin quartos on the extinct species. I do not doubt that Sir E. Lytton Bulwer had me in his mind when he introduced in one of his novels a Professor Long, who had written two huge volumes on limpets.

Although I was employed during eight years on this work, yet I record in my diary that about two years out of this time was lost by illness. On this account I went in 1848 for some months to Malvern for hydropathic treatment, which did me much good, so that on my return home I was able to resume work. So much was I out of health that when my dear father died on November 13th, 1848, I was unable to attend his funeral or to act as one of his executors.

My work on the Cirripedia possesses, I think, considerable value, as besides describing several new and remarkable forms, I made out the homologies of the various parts—I discovered the cementing apparatus, though I blundered dreadfully about the cement glands —and lastly I proved the existence in certain genera of minute males complemental to and parasitic on the hermaphrodites. This latter discovery has at last been fully confirmed ; though at one time a German writer was pleased to attribute the whole account to my fertile imagination. The Cirripedes form a highly varying and difficult group of species to class ; and my work was of considerable use to me, when I had to discuss in the 'Origin of Species' the principles of a natural classification. Nevertheless, I doubt whether

* Published by the Ray Society.

the work was worth the consumption of so much time.

From September 1854 I devoted my whole time to arranging my huge pile of notes, to observing, and to experimenting in relation to the transmutation of species. During the voyage of the *Beagle* I had been deeply impressed by discovering in the Pampean formation great fossil animals covered with armour like that on the existing armadillos; secondly, by the manner in which closely allied animals replace one another in proceeding southwards over the Continent; and thirdly, by the South American character of most of the productions of the Galapagos archipelago, and more especially by the manner in which they differ slightly on each island of the group; none of the islands appearing to be very ancient in a geological sense.

It was evident that such facts as these, as well as many others, could only be explained on the supposition that species gradually become modified; and the subject haunted me. But it was equally evident that neither the action of the surrounding conditions, nor the will of the organisms (especially in the case of plants) could account for the innumerable cases in which organisms of every kind are beautifully adapted to their habits of life—for instance, a woodpecker or a tree-frog to climb trees, or a seed for dispersal by hooks or plumes. I had always been much struck by such adaptations, and until these could be explained it seemed to me almost useless to endeavour to prove by indirect evidence that species have been modified.

After my return to England it appeared to me that by following the example of Lyell in Geology, and by collecting all facts which bore in any way on the variation of animals and plants under domestication and nature, some light might perhaps be thrown on the whole subject. My first note-book was opened in July 1837. I worked on true Baconian principles, and without any theory collected facts on a wholesale scale, more especially with respect to domesticated productions, by printed enquiries, by conversation with skilful breeders and gardeners, and by extensive reading. When I see the list of books of all kinds which I read and abstracted, including whole series of Journals and Transactions, I am surprised at my industry. I soon perceived that selection was the keystone of man's success in making useful races of animals and plants. But how selection could be applied to organisms living in a state of nature remained for some time a mystery to me.

In October 1838, that is, fifteen months after I had begun my systematic enquiry, I happened to read for amusement 'Malthus on Population,' and being well prepared to appreciate the struggle for existence which everywhere goes on from long-continued observation of the habits of animals and plants, it at once struck me that under these circumstances favourable variations would tend to be preserved, and unfavourable ones to be destroyed. The result of this would be the formation of new species. Here then I had at last got a theory by which to work ; but I was so anxious to avoid prejudice, that I determined not for some time to write even the briefest sketch of it. In June 1842 I first allowed myself the satisfaction of writing a very brief abstract of my theory in pencil in 35 pages ; and this was enlarged during the summer of 1844 into one of 230 pages, which I had fairly copied out and still possess.

But at that time I overlooked one problem of great importance ; and it is astonishing to me, except on the principle of Columbus and his egg, how I could have overlooked it and its solution. This problem is the tendency in organic beings descended from the same stock to diverge in character as they become modified. That they have diverged greatly is obvious from the manner in which species of all kinds can be classed under genera, genera under families, families under sub-orders and so forth ; and I can remember the very spot in the road, whilst in my carriage, when to my joy the solution occurred to me ; and this was long after I had come to Down. The solution, as I believe, is that the modified offspring of all dominant and increasing forms tend to become adapted to many and highly diversified places in the economy of nature.

Early in 1856 Lyell advised me to write out my views pretty fully, and I began at once to do so on a scale three or four times as extensive as that which was afterwards followed in my ' Origin of Species ; ' yet it was only an abstract of the materials which I had collected, and I got through about half the work on this scale. But my plans were overthrown, for early in the summer of 1858 Mr. Wallace, who was then in the Malay archipelago, sent me an essay " On the Tendency of Varieties to depart indefinitely from the Original Type ; " and this essay contained exactly the same theory as mine. Mr. Wallace expressed the wish that if I thought well of his essay, I should send it to Lyell for perusal.

The circumstances under which I consented at the request of Lyell and Hooker to allow of an abstract from my MS., together with a letter to Asa Gray, dated September 5, 1857, to be published at the same time with Wallace's Essay, are given in the ' Journal of the Proceedings of the Linnean Society,' 1858, p. 45. I was at first very unwilling to consent, as I thought Mr. Wallace might consider my doing so unjustifiable, for I did not then know how generous and noble was his disposition. The extract from my MS. and the letter to Asa Gray had neither been intended for publication, and were badly written. Mr. Wallace's essay, on the other hand, was admirably expressed and quite clear. Nevertheless, our joint productions excited very little attention, and the only published notice of them which I can remember was by Professor Haughton of Dublin, whose verdict was that all that was new in them was false, and what was true was old. This shows how necessary it is that any new view should be explained at considerable length in order to arouse public attention.

In September 1858 I set to work by the strong advice of Lyell and Hooker to prepare a volume on the transmutation of species, but was often interrupted by ill-health, and short visits to Dr. Lane's delightful

hydropathic establishment at Moor Park. I abstracted the MS. begun on a much larger scale in
1856, and completed the volume on the same reduced scale. It cost me thirteen months and ten
days' hard labour. It was published under the title of
the 'Origin of Species,' in November 1859. Though
considerably added to and corrected in the later
editions, it has remained substantially the same
book.

It is no doubt the chief work of my life. It was
from the first highly successful. The first small
edition of 1250 copies was sold on the day of publication, and a second edition of 3000 copies soon afterwards. Sixteen thousand copies have now (1876)
been sold in England; and considering how stiff a
book it is, this is a large sale. It has been translated
into almost every European tongue, even into such
languages as Spanish, Bohemian, Polish, and Russian.
It has also, according to Miss Bird, been translated
into Japanese,* and is there much studied. Even an
essay in Hebrew has appeared on it, showing that the
theory is contained in the Old Testament! The
reviews were very numerous; for some time I collected all that appeared on the 'Origin' and on my
related books, and these amount (excluding newspaper reviews) to 265; but after a time I gave up the
attempt in despair. Many separate essays and books
on the subject have appeared; and in Germany a
catalogue or bibliography on " Darwinismus " has
appeared every year or two.

* Miss Bird is mistaken, as I learn from Prof. Mitsukuri.—F.D.

The success of the 'Origin' may, I think, be
attributed in large part to my having long before
written two condensed sketches, and to my having
finally abstracted a much larger manuscript, which was
itself an abstract. By this means I was enabled to
select the more striking facts and conclusions. I had,
also, during many years followed a golden rule,
namely, that whenever a published fact, a new observation or thought came across me, which was opposed
to my general results, to make a memorandum of it
without fail and at once; for I had found by experience that such facts and thoughts were far more
apt to escape from the memory than favourable ones.
Owing to this habit, very few objections were raised
against my views which I had not at least noticed and
attempted to answer.

It has sometimes been said that the success of the
'Origin' proved "that the subject was in the air," or
"that men's minds were prepared for it." I do not
think that this is strictly true, for I occasionally
sounded not a few naturalists, and never happened to
come across a single one who seemed to doubt about
the permanence of species. Even Lyell and Hooker,
though they would listen with interest to me, never
seemed to agree. I tried once or twice to explain to
able men what I meant by Natural Selection, but
signally failed. What I believe was strictly true is
that innumerable well-observed facts were stored in the
minds of naturalists ready to take their proper places
as soon as any theory which would receive them was
sufficiently explained. Another element in the success

of the book was its moderate size ; and this I owe to
the appearance of Mr. Wallace's essay; had I published on the scale in which I began to write in 1856,
the book would have been four or five times as large
as the 'Origin,' and very few would have had the
patience to read it.

I gained much by my delay in publishing from
about 1839, when the theory was clearly conceived, to
1859 ; and I lost nothing by it, for I cared very little
whether men attributed most originality to me or
Wallace ; and his essay no doubt aided in the reception
of the theory.   I was forestalled in only one important
point, which my vanity has always made me regret,
namely, the explanation by means of the Glacial
period of the presence of the same species of plants
and of some few animals on distant mountain summits
and in the arctic regions.   This view pleased me so
much that I wrote it out in extenso, and I believe that
it was read by Hooker some years before E. Forbes
published his celebrated memoir * on the subject.   In
the very few points in which we differed, I still think
that I was in the right.   I have never, of course,
alluded in print to my having independently worked
out this view. '

Hardly any point gave me so much satisfaction
when I was at work on the 'Origin,' as the explanation of the wide difference in many classes between
the embryo and the adult animal, and of the close
resemblance of the embryos within the same class.
No notice of this point was taken, as far as I re-

* ' Geolog. Survey Mem.,' 1846.

member, in the early reviews of the 'Origin,' and I recollect expressing my surprise on this head in a letter to Asa - Gray. Within late years several reviewers have given the whole credit to Fritz Müller and Häckel, who undoubtedly have worked it out much more fully, and in some respects more correctly than I did. I had materials for a whole chapter on the subject, and I ought to have made the discussion longer; for it is clear that I failed to impress my readers ; and he who succeeds in doing so deserves, in my opinion, all the credit.

This leads me to remark that I have almost always been treated honestly by my reviewers, passing over those without scientific knowledge as not worthy of notice. My views have often been grossly misrepresented, bitterly opposed and ridiculed, but this has been generally done as, I believe, in good faith. On the whole I do not doubt that my works have been over and over again greatly overpraised. I rejoice that I have avoided controversies, and this I owe to Lyell, who many years ago, in reference to my geological works, strongly advised me never to get entangled in a controversy, as it rarely did any good and caused a miserable loss of time and temper.

Whenever I have found out that I have blundered, or that my work has been imperfect, and when I have been contemptuously criticised, and even when I have been overpraised, so that I have felt mortified, it has been my greatest comfort to say hundreds of times to myself that " I have worked as hard and as well as I could, and no man can do more than this." I remember when in Good Success Bay, in Tierra del Fuego, thinking (and, I believe, that I wrote home to the effect) that I could not employ my life better than in adding a little to Natural Science. This I have done to the best of my abilities, and critics may say what they like, but they cannot destroy this conviction.

During the two last months of 1859 I was fully occupied in preparing a second edition of the 'Origin,' and by an enormous correspondence. On January 1st, 1860, I began arranging my notes for my work on the 'Variation of Animals and Plants under Domestication;' but it was not published until the beginning of 1868; the delay having been caused partly by frequent illnesses, one of which lasted seven months, and partly by being tempted to publish on other subjects which at the time interested me more.

On May 15th, 1862, my little book on the 'Fertilisation of Orchids,' which cost me ten months' work, was published : most of the facts had been slowly accumulated during several previous years. During the summer of 1839, and, I believe, during the previous summer, I was led to attend to the cross-fertilisation of flowers by the aid of insects, from having come to the conclusion in my speculations on the origin of species, that crossing played an important part in keeping specific forms constant. I attended to the subject more or less during every subsequent summer; and my interest in it was greatly enhanced by having procured and read in November 1841, through the advice of Robert Brown, a copy of C. K. Sprengel's wonderful book, 'Das entdeckte Geheimniss

der Natur.' For some years before 1862 I had
specially attended to the fertilisation of our British
orchids ; and it seemed to me the best plan to prepare
as complete a treatise on this group of plants as well
as I could, rather than to utilise the great mass of
matter which I had slowly collected with respect to
other plants.

My resolve proved a wise one ; for since the appearance of my book, a surprising number of papers
and separate works on the fertilisation of all kinds of
flowers have appeared ; and these are far better done
than I could possibly have effected. The merits of
poor old Sprengel, so long overlooked, are now fully
recognised many years after his death.

During the same year I published in the ' Journal of
the Linnean Society' a paper "On the Two Forms, or
Dimorphic Condition of Primula," and during the next
five years, five other papers on dimorphic and trimorphic plants. I do not think anything in my
scientific life has given me so much satisfaction as
making out the meaning of the structure of these
plants. I had noticed in 1838 or 1839 the dimorphism of *Linum flavum*, and had at first thought that
it was merely a case of unmeaning variability. But
on examining the common species of Primula I found
that the two forms were much too regular and constant
to be thus viewed. I therefore became almost convinced that the common cowslip and primrose were
on the high-road to become diœcious ;—that the short
pistil in the one form, and the short stamens in the
other form were tending towards abortion. The plants

were therefore subjected under this point of view to
trial; but as soon as the flowers with short pistils
fertilised with pollen from the short stamens, were
found to yield more seeds than any other of the four
possible unions, the abortion-theory was knocked on
the head.   After some additional experiment, it became evident that the two forms, though both were
perfect hermaphrodites, bore almost the same relation
to one another as do the two sexes of an ordinary
animal.   With Lythrum we have the still more wonderful case of three forms standing in a similar relation
to one another.   I afterwards found that the offspring
from the union of two plants belonging to the same
forms presented a close and curious analogy with
hybrids from the union of two distinct species.

In the autumn of 1864 I finished a long paper on
'Climbing Plants,' and sent it to the Linnean Society.
The writing of this paper cost me four months; but I
was so unwell when I received the proof-sheets that
I was forced to leave them very badly and often ob-
scurely expressed.   The paper was little noticed, but
when in 1875 it was corrected and published as a
separate book it sold well.   I was led to take up this
subject by reading a short paper by Asa Gray, published in 1858.   He sent me seeds, and on raising
some plants I was so much fascinated and perplexed
by the revolving movements of the tendrils and stems,
which movements are really very simple, though appearing at first sight very complex, that I procured
various other kinds of climbing plants, and studied the
whole subject.   I was all the more attracted to it,

from not being at all satisfied with the explanation
which Henslow gave us in his lectures, about twining
plants, namely, that they had a natural tendency to
grow up in a spire. This explanation proved quite
erroneous. Some of the adaptations displayed by
Climbing Plants are as beautiful as those of Orchids
for ensuring cross-fertilisation.

My ' Variation of Animals and Plants under Domestication' was begun, as already stated, in the beginning of 1860, but was not published until the
beginning of 1868. It was a big book, and cost me
four years and two months' hard labour. It gives all
my observations and an immense number of facts
collected from various sources, about our domestic
productions. In the second volume the causes and
laws of variation, inheritance, &c., are discussed as far
as our present state of knowledge permits. Towards
the end of the work I give my well-abused hypothesis
of Pangenesis. An unverified hypothesis is of little
or no value ; but if any one should hereafter be led to
make observations by which some such hypothesis
could be established, I shall have done good service,
as an astonishing number of isolated facts can be thus
connected together and rendered intelligible. In 1875
a second and largely corrected edition, which cost me
a good deal of labour, was brought out.

My ' Descent of Man ' was published in February
1871. As soon as I had become, in the year 1837 or
1838, convinced that species were mutable productions,
I could not avoid the belief that man must come under
the same law. Accordingly I collected notes on the

subject for my own satisfaction, and not for a long
time with any intention of publishing. Although in
the 'Origin of Species' the derivation of any particular
species is never discussed, yet I thought it best, in
order that no honourable man should accuse me of
concealing my views, to add that by the work "light
would be thrown on the origin of man and his history."
It would have been useless and injurious to the success
of the book to have paraded, without giving any
evidence, my conviction with respect to his origin.

But when I found that many naturalists fully accepted the doctrine of the evolution of species, it
seemed to me advisable to work up such notes as I
possessed, and to publish a special treatise on the
origin of man. I was the more glad to do so, as it
gave me an opportunity of fully discussing sexual
selection—a subject which had always greatly interested me. This subject, and that of the variation of
our domestic productions, together with the causes
and laws of variation, inheritance, and the intercrossing
of plants, are the sole subjects which I have been able
to write about in full, so as to use all the materials
which I have collected. The 'Descent of Man' took
me three years to write, but then as usual some of
this time was lost by ill-health, and some was consumed
by preparing new editions and other minor works.
A second and largely corrected edition of the 'Descent'
appeared in 1874.

My book on the 'Expression of the Emotions in
Men and Animals' was published in the autumn of
1872. I had intended to give only a chapter on the

subject in the 'Descent of Man,' but as soon as I
began to put my notes together, I saw that it would
require a separate treatise. My first child was born on December 27th, 1839, and
I at once commenced to make notes on the first dawn
of the various expressions which he exhibited, for I
felt convinced, even at this early period, that the most
complex and fine shades of expression must all have
had a gradual and natural origin. During the summer
of the following year, 1840, I read Sir C. Bell's admirable work on expression, and this greatly increased
the interest which I felt in the subject, though I could
not at all agree with his belief that various muscles
had been specially created for the sake of expression.
From this time forward I occasionally attended to the
subject, both with respect to man and our domesticated
animals. My book sold largely; 5267 copies having
been disposed of on the day of publication.

In the summer of 1860 I was idling and resting
near Hartfield, where two species of Drosera abound ;
and I noticed that numerous insects had been en-
trapped by the leaves. I carried home some plants,
and on giving them insects saw the movements of the
tentacles, and this made me think it probable that the
insects were caught for some special purpose. Fortunately a crucial test occurred to me, that of placing a
large number of leaves in various nitrogenous and
non-nitrogenous fluids of equal density ; and as soon
as I found that the former alone excited energetic
movements, it was obvious that here was a fine new
field for investigation.

During subsequent years, whenever I had leisure, I
pursued my experiments, and my book on 'Insectivorous Plants' was published in July 1875—that is
sixteen years after my first observations. The delay
in this case, as with all my other books, has been a
great advantage to me ; for a man after a long interval
can criticise his own work, almost as well as if it were
that of another person. The fact that a plant should
secrete, when properly excited, a fluid containing an
acid and ferment, closely analogous to the digestive
fluid of an animal, was certainly a remarkable discovery.

During this autumn of 1876 I shall publish on the
'Effects of Cross and Self-Fertilisation in the Vegetable Kingdom.' This book will form a complement
to that on the 'Fertilisation of Orchids,' in which I
showed how perfect were the means for cross-fertilisation, and here I shall show how important are the
results. I was led to make, during eleven years, the
numerous experiments recorded in this volume, by a
mere accidental observation ; and indeed it required
the accident to be repeated before my attention was
thoroughly aroused to the remarkable fact that seedlings of self-fertilised parentage are inferior, even in
the first generation, in height and vigour to seedlings
of cross-fertilised parentage. I hope also to republish
a revised edition of my book on Orchids, and hereafter
my papers on dimorphic and trimorphic plants, together
with some additional observations on allied points
which I never have had time to arrange. My strength
will then probably be exhausted, and I shall be ready
to exclaim " Nunc dimittis."

Written May 1st, 1881.—' The Effects of Cross and Self-Fertilisation' was published in the autumn of 1876 ; and the results there arrived at explain, as I believe, the endless and wonderful contrivances for the trans-portal of pollen from one plant to another of the same species. I now believe, however, chiefly from the observations of Hermann Müller, that I ought to have insisted more strongly than I did on the many adapta-tions for self-fertilisation ; though I was well aware of many such adaptations. A much enlarged edition of my ' Fertilisation of Orchids' was published in 1877.

In this same year ' The Different Forms of Flowers, &c.,' appeared, and in 1880 a second edition. This book consists chiefly of the several papers on Hetero-styled flowers originally published by the Linnean Society, corrected, with much new matter added, together with observations on some other cases in which the same plant bears two kinds of flowers. As before remarked, no little discovery of mine ever gave me so much pleasure as the making out the meaning of heterostyled flowers. The results of crossing such flowers in an illegitimate manner, I believe to be very important, as bearing on the sterility of hybrids ; although these results have been noticed by only a few persons.

In 1879, I had a translation of Dr. Ernst Krause's ' Life of Erasmus Darwin' published, and I added a sketch of his character and habits from material in my possession. Many persons have been much inte-rested by this little life, and I am surprised that only 800 or 900 copies were sold.

In 1880 I published, with [my son] Frank's assistance, our ' Power of Movement in Plants.' This was a tough piece of work. The book bears somewhat the same relation to my little book on ' Climbing Plants,' which ' Cross-Fertilisation ' did to the ' Fertilisation of Orchids ; ' for in accordance with the principle of evolution it was impossible to account for climbing plants having been developed in so many widely different groups unless all kinds of plants possess some slight power of movement of an analogous kind. This I proved to be the case ; and I was further led to a rather wide generalisation, viz. that the great and important classes of movements, excited by light, the attraction of gravity, &c., are all modified forms of the fundamental movement of circumnutation. It has always pleased me to exalt plants in the scale of organised beings ; and I therefore felt an especial pleasure in showing how many and what admirably well adapted movements the tip of a root possesses.

I have now (May 1, 1881) sent to the printers the MS. of a little book on ' The Formation of Vegetable Mould, through the Action of Worms.' This is a subject of but small importance ; and I know not whether it will interest any readers,* but it has interested me. It is the completion of a short paper read before the Geological Society more than forty years ago, and has revived old geological thoughts.

I have now mentioned all the books which I have published, and these have been the milestones in my

---

* Between November 1881 and February 1884, 8500 copies have been sold.

life, so that little remains to be said. I am not conscious of any change in my mind during the last thirty years, excepting in one point presently to be mentioned; nor, indeed, could any change have been expected unless one of general deterioration. But my father lived to his eighty-third year with his mind as lively as ever it was, and all his faculties undimmed; and I hope that I may die before my mind fails to a sensible extent. I think that I have become a little more skilful in guessing right explanations and in devising experimental tests; but this may probably be the result of mere practice, and of a larger store of knowledge. I have as much difficulty as ever in expressing myself clearly and concisely; and this difficulty has caused me a very great loss of time; but it has had the compensating advantage of forcing me to think long and intently about every sentence, and thus I have been led to see errors in reasoning and in my own observations or those of others.

There seems to be a sort of fatality in my mind leading me to put at first my statement or proposition in a wrong or awkward form. Formerly I used to think about my sentences before writing them down; but for several years I have found that it saves time to scribble in a vile hand whole pages as quickly as I possibly can, contracting half the words; and then correct deliberately. Sentences thus scribbled down are often better ones than I could have written deliberately.

Having said thus much about my manner of writing, I will add that with my large books I spend a good deal of time over the general arrangement of the matter. I first make the rudest outline in two or three pages, and then a larger one in several pages, a few words or one word standing for a whole discussion or series of facts. Each one of these headings is again enlarged and often transferred before I begin to write *in extenso*. As in several of my books facts observed by others have been very extensively used, and as I have always had several quite distinct subjects in hand at the same time, I may mention that I keep from thirty to forty large portfolios, in cabinets with labelled shelves, into which I can at once put a detached reference or memorandum. I have bought many books, and at their ends I make an index of all the facts that concern my work ; or, if the book is not my own, write out a separate abstract, and of such abstracts I have a large drawer full. Before beginning on any subject I look to all the short indexes and make a general and classified index, and by taking the one or more proper portfolios I have all the information collected during my life ready for use.

I have said that in one respect my mind has changed during the last twenty or thirty years. Up to the age of thirty, or beyond it, poetry of many kinds, such as the works of Milton, Gray, Byron, Wordsworth, Coleridge, and Shelley, gave me great pleasure, and even as a schoolboy I took intense delight in Shakespeare, especially in the historical plays. I have also said that formerly pictures gave me considerable, and music very great delight. But now for many years I cannot endure to read a line of poetry : I have tried

lately to read Shakespeare, and found it so intolerably
dull that it nauseated me. I have also almost lost my
taste for pictures or music. Music generally sets me
thinking too energetically on what I have been at
work on, instead of giving me pleasure. I retain
some taste for fine scenery, but it does not cause me
the exquisite delight which it formerly did. On the
other hand, novels which are works of the imagination,
though not of a very high order, have been for years
a wonderful relief and pleasure to me, and I often
bless all novelists. A surprising number have been
read aloud to me, and I like all if moderately good,
and if they do not end unhappily—against which a
law ought to be passed. A novel, according to my
taste, does not come into the first class unless it contains some person whom one can thoroughly love,
and if a pretty woman all the better.

This curious and lamentable loss of the higher
æsthetic tastes is all the odder, as books on history,
biographies, and travels (independently of any scientific facts which they may contain), and essays on all
sorts of subjects interest me as much as ever they did.
My mind seems to have become a kind of machine for
grinding general laws out of large collections of facts,
but why this should have caused the atrophy of that
part of the brain alone, on which the higher tastes
depend, I cannot conceive. A man with a. mind
more highly organised or better constituted than
mine, would not, I suppose, have thus suffered ; and
if I had to live my life again, I would have made a
rule to read some poetry and listen to some music at

least once every week; for perhaps the parts of my brain now atrophied would thus have been kept active through use. The loss of these tastes is a loss of happiness, and may possibly be injurious to the intellect, and more probably to the moral character, by enfeebling the emotional part of our nature.

My books have sold largely in England, have been translated into many languages, and passed through several editions in foreign countries. I have heard it said that the success of a work abroad is the best test of its enduring value. I doubt whether this is at all trustworthy; but judged by this standard my name ought to last for a few years. Therefore it may be worth while to try to analyse the mental qualities and the conditions on which my success has depended; though I am aware that no man can do this correctly.

I have no great quickness of apprehension or wit which is so remarkable in some clever men, for instance, Huxley. I am therefore a poor critic: a paper or book, when first read, generally excites my admiration, and it is only after considerable reflection that I perceive the weak points. My power to follow a long and purely abstract train of thought is very limited; and therefore I could never have succeeded with metaphysics or mathematics. My memory is extensive, yet hazy: it suffices to make me cautious by vaguely telling me that I have observed or read something opposed to the conclusion which I am drawing, or on the other hand in favour of it; and after a time I can generally recollect where to search for my authority. So poor in one sense is my memory,

that I have never been able to remember for more
than a few days a single date or a line of poetry.

Some of my critics have said, " Oh, he is a good
observer, but he has no power of reasoning !" I do
not think that this can be true, for the ' Origin of
Species' is one long argument from the beginning to
the end, and it has convinced not a few able men.
No one could have written it without having some
power of reasoning. I have a fair share of invention,
and of common sense or judgment, such as every
fairly successful lawyer or doctor must have, but not,
I believe, in any higher degree.

On the favourable side of the balance, I think that
I am superior to the common run of men in noticing
things which easily escape attention, and in observing
them carefully. My industry has been nearly as great
as it could have been in the observation and collection
of facts. What is far more important, my love of
natural science has been steady and ardent.

This pure love has, however, been much aided by
the ambition to be esteemed by my fellow naturalists.
From my early youth I have had the strongest desire
to understand or explain whatever I observed,—that
is, to group all facts under some general laws. These
causes combined have given me the patience to reflect
or ponder for any number of years over any unex-
plained problem. As far as I can judge, I am not
apt to follow blindly the lead of other men. I have
steadily endeavoured to keep my mind free so as to
give up any hypothesis, however much beloved (and I
cannot resist forming one on every subject), as soon

as facts are shown to be opposed to it. Indeed, I
have had no choice but to act in this manner, for with
the exception of the Coral Reefs, I cannot remember a
single first-formed hypothesis which had not after a
time to be given up or greatly modified. This has
naturally led me to distrust greatly deductive reasoning in the mixed sciences. On the other hand, I am
not very sceptical,—a frame of mind which I believe
to be injurious to the progress of science. A good
deal of scepticism in a scientific man is advisable to
avoid much loss of time, for I have met with not a few
men, who, I feel sure, have often thus been deterred
from experiment or observations, which would have
proved directly or indirectly serviceable.

In illustration, I will give the oddest case which I
have known. A gentleman (who, as I afterwards
heard, is a good local botanist) wrote to me from the
Eastern counties that the seeds or beans of the common field-bean had this year everywhere grown on the
wrong side of the pod. I wrote back, asking for
further information, as I did not understand what was
meant; but I did not receive any answer for a very
long time. I then saw in two newspapers, one published in Kent and the other in Yorkshire, paragraphs
stating that it was a most remarkable fact that "the
beans this year had all grown on the wrong side." So
I thought there must be some foundation for so general
a statement. Accordingly, I went to my gardener, an
old Kentish man, and asked him whether he had
heard anything about it, and he answered, "Oh, no,
sir, it must be a mistake, for the beans grow on the

wrong side only on leap-year, and this is not leap-year." I then asked him how they grew in common years and how on leap-years, but soon found that he knew absolutely nothing of how they grew at any time, but he stuck to his belief. After a time I heard from my first informant, who, with many apologies, said that he should not have written to me had he not heard the statement from several intelligent farmers; but that he had since spoken again to every one of them, and not one knew in the least what he had himself meant. So that here a belief—if indeed a statement with no definite idea attached to it can be called a belief—had spread over almost the whole of England without any vestige of evidence.

I have known in the course of my life only three intentionally falsified statements, and one of these may have been a hoax (and there have been several scientific hoaxes) which, however, took in an American Agricultural Journal. It related to the formation in Holland of a new breed of oxen by the crossing of distinct species of Bos (some of which I happen to know are sterile together), and the author had the impudence to state that he had corresponded with me, and that I had been deeply impressed with the importance of his result. The article was sent to me by the editor of an English Agricultural Journal, asking for my opinion before republishing it.

A second case was an account of several varieties, raised by the author from several species of Primula, which had spontaneously yielded a full complement of seed, although the parent plants had been carefully protected from the access of insects. This account was published before I had discovered the meaning of heterostylism, and the whole statement must have been fraudulent, or there was neglect in excluding insects so gross as to be scarcely credible.

The third case was more curious :   Mr. Huth published in his book on 'Consanguineous Marriage' some long extracts from a Belgian author, who stated that he had interbred rabbits in the closest manner for very many generations, without the least injurious effects. The account was published in a most respectable Journal, that of the Royal Society of Belgium ; but I could not avoid feeling doubts—I hardly know why, except that there were no accidents of any kind, and my experience in breeding animals made me think this very improbable.

So with much hesitation I wrote to Professor Van Beneden, asking him whether the author was a trust-worthy man.   I soon heard in answer that the Society had been greatly shocked by discovering that the whole account was a fraud.*   The writer had been publicly challenged in the Journal to say where he had resided and kept his large stock of rabbits while carrying on his experiments, which must have consumed several years, and no answer could be extracted from him.

My habits are methodical, and this has been of not

* The falseness of the published statements on which Mr. Huth relied has been pointed out by him-self in a slip inserted in all the copies of his book which then remained unsold.

a little use for my particular line of work. Lastly, I have had ample leisure from not having to earn my own bread. Even ill-health, though it has annihilated several years of my life, has saved me from the distractions of society and amusement.

Therefore my success as a man of science, whatever this may have amounted to, has been determined, as far as I can judge, by complex and diversified mental qualities and conditions. Of these, the most important have been—the love of science—unbounded patience in long reflecting over any subject—industry in observing and collecting facts—and a fair share of invention as well as of common sense. With such moderate abilities as I possess, it is truly surprising that I should have influenced to a considerable extent the belief of scientific men on some important points.

THE STUDY AT DOWN.*

CHAPTER III.

REMINISCENCES OF MY FATHER'S EVERYDAY LIFE.

IT is my wish in the present chapter to give some idea of my
father's everyday life. It has seemed to me that I might
carry out this object in the form of a rough sketch of a day's
life at Down, interspersed with such recollections as are called
up by the record. Many of these recollections, which have
a meaning for those who knew my father, will seem colourless
or trifling to strangers. Nevertheless, I give them in the
hope that they may help to preserve that impression of his
personality which remains on the minds of those who knew

* From the 'Century Magazine,' January 1883.

and loved him—an impression at once so vivid and so untranslatable into words.

Of his personal appearance (in these days of multiplied photographs) it is hardly necessary to say much. He was about six feet in height, but scarcely looked so tall, as he stooped a good deal ; in later days he yielded to the stoop ; but I can remember seeing him long ago swinging his arms back to open out his chest, and holding himself upright with a jerk. He gave one the idea that he had been active rather than strong ; his shoulders were not broad for his height, though certainly not narrow. As a young man he must have had much endurance, for on one of the shore excursions from the *Beagle*, when all were suffering from want of water, he was one of the two who were better able than the rest to struggle on in search of it. As a boy he was active, and could jump a bar placed at the height of the " Adam's apple " in his neck.

He walked with a swinging action, using a stick heavily shod with iron, which he struck loudly against the ground, producing as he went round the " Sand-walk " at Down, a rhythmical click which is with all of us a very distinct remembrance. As he returned from the midday walk, often carrying the waterproof or cloak which had proved too hot, one could see that the swinging step was kept up by something of an effort. Indoors his step was often slow and laboured, and as he went upstairs in the afternoon he might be heard mounting the stairs with a heavy footfall, as if each step were an effort. When interested in his work he moved about quickly and easily enough, and often in the middle of dictating he went eagerly into the hall to get a pinch of snuff, leaving the study door open, and calling out the last words of his sentence as he went. Indoors he sometimes used an oak stick like a little alpenstock, and this was a sign that he felt giddiness.

In spite of his strength and activity, I think he must always have had a clumsiness of movement. He was naturally awkward with his hands, and was unable to draw at all well.* This he always regretted much, and he frequently urged the paramount necessity of a young naturalist making himself a good draughtsman.

He could dissect well under the simple microscope, but I think it was by dint of his great patience and carefulness. It was characteristic of him that he thought many little bits of skilful dissection something almost superhuman. He used to speak with admiration of the skill with which he saw Newport dissect a humble bee, getting out the nervous system with a few cuts of a fine pair of scissors, held, as my father used to show, with the elbow raised, and in an attitude which certainly would render great steadiness necessary. He used to consider cutting sections a great feat, and in the last year of his life, with wonderful energy, took the pains to learn to cut sections of roots and leaves. His hand was not steady enough to hold the object to be cut, and he employed a common microtome, in which the pith for holding the object was clamped, and the razor slid on a glass surface in making the sections. He used to laugh at himself, and at his own skill in section-cutting, at which he would say he was "speechless with admiration." On the other hand, he must have had accuracy of eye and power of co-ordinating his movements, since he was a good shot with a gun as a young man, and as a boy was skilful in throwing. He once killed a hare sitting in the flower-garden at Shrewsbury by throwing a marble at it, and, as a man, he once killed a cross-beak with a stone. He was so unhappy at having uselessly killed the cross-beak that he did not mention it for years, and then explained that he should never have thrown at it if he had not felt sure that his old skill had gone from him.

When walking he had a fidgeting movement with his

* The figure representing the aggregated cell-contents in 'Insectivorous Plants' was drawn by him.

fingers, which he has described in one of his books as the habit of an old man. When he sat still he often took hold of one wrist with the other hand; he sat with his legs crossed, and from being so thin they could be crossed very far, as may be seen in one of the photographs. He had his chair in the study and in the drawing-room raised so as to be much higher than ordinary chairs; this was done because sitting on a low or even an ordinary chair caused him some discomfort. We used to laugh at him for making his tall drawing-room chair still higher by putting footstools on it, and then neutralising the result by resting his feet on another chair.

His beard was full and almost untrimmed, the hair being grey and white, fine rather than coarse, and wavy or frizzled. His moustache was somewhat disfigured by being cut short and square across. He became very bald, having only a fringe of dark hair behind.

His face was ruddy in colour, and this perhaps made people think him less of an invalid than he was. He wrote to Dr. Hooker (June 13, 1849), "Every one tells me that I look quite blooming and beautiful; and most think I am shamming, but you have never been one of those." And it must be remembered that at this time he was miserably ill, far worse than in later years. His eyes were bluish grey under deep overhanging brows, with thick bushy projecting eyebrows. His high forehead was much wrinkled, but otherwise his face was not much marked or lined. His expression showed no signs of the continual discomfort he suffered.

When he was excited with pleasant talk his whole manner was wonderfully bright and animated, and his face shared to the full in the general animation. His laugh was a free and sounding peal, like that of a man who gives himself sympathetically and with enjoyment to the person and the thing which have amused him. He often used some sort of gesture with his laugh, lifting up his hands or bringing one down with a slap. I think, generally speaking, he was given to gesture, and often used his hands in explaining anything (*e.g.* the fertilisation of a flower) in a way that seemed rather an aid to himself than to the listener. He did this on occasions when most people would illustrate their explanations by means of a rough pencil sketch.

He wore dark clothes, of a loose and easy fit. Of late years he gave up the tall hat even in London, and wore a soft black one in winter, and a big straw hat in summer. His usual out-of-doors dress was the short cloak in which Elliot and Fry's photograph represents him leaning against the pillar of the verandah. Two peculiarities of his indoor dress were that he almost always wore a shawl over his shoulders, and that he had great loose cloth boots lined with fur which he could slip on over his indoor shoes. Like most delicate people he suffered from heat as well as from chilliness; it was as if he could not hit the balance between too hot and too cold; often a mental cause would make him too hot, so that he would take off his coat if anything went wrong in the course of his work.

He rose early, chiefly because he could not lie in bed, and I think he would have liked to get up earlier than he did. He took a short turn before breakfast, a habit which began when he went for the first time to a water-cure esta-blishment. This habit he kept up till almost the end of his life. I used, as a little boy, to like going out with him, and I have a vague sense of the red of the winter sunrise, and a recollection of the pleasant companionship, and a certain honour and glory in it. He used to delight me as a boy by telling me how, in still earlier walks, on dark winter mornings, he had once or twice met foxes trotting home at the dawning.

After breakfasting alone about 7.45, he went to work at once, considering the $1\frac{1}{2}$ hour between 8 and 9.30 one of his best working times. At 9.30 he came into the drawing-room for his letters—rejoicing if the post was a light one and being sometimes much worried if it was not. He would then hear any family letters read aloud as he lay on the sofa.

The reading aloud, which also included part of a novel, lasted till about half-past ten, when he went back to work till twelve or a quarter past. By this time he considered his day's work over, and would often say, in a satisfied voice, "*I've* done a good day's work." He then went out of doors whether it was wet or fine ; Polly, his white terrier, went with him in fair weather, but in rain she refused or might be seen hesitating in the verandah, with a mixed expression of disgust and shame at her own want of courage ; generally, however, her conscience carried the day, and as soon as he was evidently gone she could not bear to stay behind.

My father was always fond of dogs, and as a young man had the power of stealing away the affections of his sisters' pets ; at Cambridge, he won the love of his cousin W. D. Fox's dog, and this may perhaps have been the little beast which used to creep down inside his bed and sleep at the foot every night. My father had a surly dog, who was devoted to him, but unfriendly to every one else, and when he came back from the *Beagle* voyage, the dog remembered him, but in a curious way, which my father was fond of telling. He went into the yard and shouted in his old manner ; the dog rushed out and set off with him on his walk, showing no more emotion or excitement than if the same thing had happened the day before, instead of five years ago. This story is made use of in the 'Descent of Man,' 2nd Edit. p. 74.

In my memory there were only two dogs which had much connection with my father. One was a large black and white half-bred retriever, called Bob, to which we, as children, were much devoted. He was the dog of whom the story of the " hot-house face " is told in the 'Expression of the Emotions.'

But the dog most closely associated with my father was the above-mentioned Polly, a rough, white fox-terrier. She was a sharp-witted, affectionate dog ; when her master was going away on a journey, she always discovered the fact by the signs of packing going on in the study, and became low-spirited accordingly. She began, too, to be excited by seeing the study prepared for his return home. She was a cunning little creature, and used to tremble or put on an air of misery when my father passed, while she was waiting for dinner, just as if she knew that he would say (as he did often say) that "she was famishing." My father used to make her catch biscuits off her nose, and had an affectionate and mock-solemn way of explaining to her before-hand that she must "be a very good girl." She had a mark on her back where she had been burnt, and where the hair had re-grown red instead of white, and my father used to commend her for this tuft of hair as being in accordance with his theory of pangenesis ; her father had been a red bull-terrier, thus the red hair appearing after the burn showed the presence of latent red gemmules. He was delightfully tender to Polly, and never showed any impatience at the attentions she required, such as to be let in at the door, or out at the verandah window, to bark at "naughty people," a self-imposed duty she much enjoyed. She died, or rather had to be killed, a few days after his death.*

My father's midday walk generally began by a call at the greenhouse, where he looked at any germinating seeds or experimental plants which required a casual examination, but he hardly ever did any serious observing at this time. Then he went on for his constitutional—either round the "Sand-walk," or outside his own grounds in the immediate neighbourhood of the house. The "Sand-walk" was a narrow strip of land 1½ acres in extent, with a gravel-walk round it. On one side of it was a broad old shaw with fair-sized

* The basket in which she usually lay curled up near the fire in his study is faithfully represented in Mr. Parson's drawing given at the head of the chapter.

oaks in it, which made a sheltered shady walk ; the other side was separated from a neighbouring grass field by a low quickset hedge, over which you could look at what view there was, a quiet little valley losing itself in the upland country towards the edge of the Westerham hill, with hazel coppice and larch wood, the remnants of what was once a large wood, stretching away to the Westerham road. I have heard my father say that the charm of this simple little valley helped to make him settle at Down.

The Sand-walk was planted by my father with a variety of trees, such as hazel, alder, lime, hornbeam, birch, privet, and dogwood, and with a long line of hollies all down the exposed side. In earlier times he took a certain number of turns every day, and used to count them by means of a heap of flints, one of which he kicked out on the path each time he passed. Of late years I think he did not keep to any fixed number of turns, but took as many as he felt strength for. The Sand-walk was our play-ground as children, and here we continually saw my father as he walked round. He liked to see what we were doing, and was ever ready to sympathize in any fun that was going on. It is curious to think how, with regard to the Sand-walk in connection with my father, my earliest recollections coincide with my latest ; it shows how unvarying his habits have been.

Sometimes when alone he stood still or walked stealthily to observe birds or beasts. It was on one of these occasions that some young squirrels ran up his back and legs, while their mother barked at them in an agony from the tree. He always found birds' nests even up to the last years of his life, and we, as children, considered that he had a special genius in this direction. In his quiet prowls he came across the less common birds, but I fancy he used to conceal it from me, as a little boy, because he observed the agony of mind which I endured at not having seen the siskin or goldfinch, or whatever it might have been. He used to tell us how, when he was creeping noiselessly along in the "Big-Woods," he came upon a fox asleep in the daytime, which was so much astonished that it took a good stare at him before it ran off. A Spitz dog which accompanied him showed no sign of excitement at the fox, and he used to end the story by wondering how the dog could have been so faint-hearted.

Another favourite place, was "Orchis Bank," above the quiet Cudham valley, where fly- and musk-orchis grew among the junipers, and Cephalanthera and Neottia under the beech boughs ; the little wood "Hangrove," just above this, he was also fond of, and here I remember his collecting grasses, when he took a fancy to make out the names of all the common kinds. He was fond of quoting the saying of one of his little boys, who, having found a grass that his father had not seen before, had it laid by his own plate during dinner, remarking, "I are an extraordinary grass-finder!"

My father much enjoyed wandering slowly in the garden with my mother or some of his children, or making one of a party, sitting out on a bench on the lawn ; he generally sat, however, on the grass, and I remember him often lying under one of the big lime-trees, with his head on the green mound at its foot. In dry summer weather, when we often sat out, the big fly-wheel of the well was commonly heard spinning round, and so the sound became associated with those pleasant days. He used to like to watch us playing at lawn-tennis, and often knocked up a stray ball for us with the curved handle of his stick.

Though he took no personal share in the management of the garden, he had great delight in the beauty of flowers— for instance, in the mass of Azaleas which generally stood in the drawing-room. I think he sometimes fused together his admiration of the structure of a flower and of its intrinsic beauty ; for instance, in the case of the big pendulous pink and white flowers of Dielytra. In the same way he had an affection, half-artistic, half-botanical, for the little blue Lobelia. In admiring flowers, he would often laugh at the dingy high-art colours, and contrast them with the bright tints of nature. I used to like to hear him admire the beauty of a flower; it was a kind of gratitude to the flower itself, and a personal love for its delicate form and colour. I seem to remember him gently touching a flower he delighted in; it was the same simple admiration that a child might have.

He could not help personifying natural things. This feeling came out in abuse as well as in praise—*e.g.* of some seedlings —"The little beggars are doing just what I don't want them to." He would speak in a half-provoked, half-admiring way of the ingenuity of a Mimosa leaf in screwing itself out of a basin of water in which he had tried to fix it. One might see the same spirit in his way of speaking of Sundew, earthworms, &c.*

Within my memory, his only outdoor recreation, besides walking, was riding, which he took to on the recommendation of Dr. Bence Jones, and we had the luck to find for him the easiest and quietest cob in the world, named "Tommy." He enjoyed these rides extremely, and devised a number of short rounds which brought him home in time for lunch. Our country is good for this purpose, owing to the number of small valleys which give a variety to what in a flat country would be a dull loop of road. He was not, I think, naturally fond of horses, nor had he a high opinion of their intelligence, and Tommy was often laughed at for the alarm he showed at passing and repassing the same heap of hedge-clippings as he went round the field. I think he used to feel surprised at himself, when he remembered how bold a rider he had been, and how utterly old age and bad health had taken away his nerve. He would say that riding prevented him thinking

---

* Cf. Leslie Stephen's 'Swift,' 1882, p. 200, where Swift's inspection of the manners and customs of servants are compared to my father's observations on worms, "The difference is," says Mr. Stephen, "that Darwin had none but kindly feelings for worms."

much more effectually than walking—that having to attend
to the horse gave him occupation sufficient to prevent any
really hard thinking. And the change of scene which it gave
him was good for spirits and health.

Unluckily, Tommy one day fell heavily with him on
Keston common. This, and an accident with another horse
upset his nerves, and he was advised to give up riding.

If I go beyond my own experience, and recall what I have
heard him say of his love for sport, &c., I can think of a good
deal, but much of it would be a repetition of what is contained in his 'Recollections.' At school he was fond of bat-
fives, and this was the only game at which he was skilful.
He was fond of his gun as quite a boy, and became a good
shot ; he used to tell how in South America he killed twenty-
three snipe in twenty-four shots. In telling the story he was
careful to add that he thought they were not quite so wild
as English snipe.

Luncheon at Down came after his midday walk ; and
here I may say a word or two about his meals generally.
He had a boy-like love of sweets, unluckily for himself, since
he was constantly forbidden to take them. He was not
particularly successful in keeping the "vows," as he called
them, which he made against eating sweets, and never considered them binding unless he made them aloud.

He drank very little wine, but enjoyed, and was revived
by, the little he did drink. He had a horror of drinking,
and constantly warned his boys that any one might be led
into drinking too much. I remember, in my innocence as a
small boy, asking him if he had been ever tipsy ; and he
answered very gravely that he was ashamed to say he had
once drunk too much at Cambridge. I was much impressed,
so that I know now the place where the question was asked.

After his lunch, he read the newspaper, lying on the sofa
in the drawing-room. I think the paper was the only non-
scientific matter which he read to himself. Everything else,

novels, travels, history, was read aloud to him. He took so wide an interest in life, that there was much to occupy him in newspapers, though he laughed at the wordiness of the debates ; reading them, I think, only in abstract. His interest in politics was considerable, but his opinion on these matters was formed rather by the way than with any serious amount of thought.

After he had read his paper, came his time for writing letters. These, as well as the MS. of his books, were written by him as he sat in a huge horse-hair chair by the fire, his paper supported on a board resting on the arms of the chair. When he had many or long letters to write, he would dictate them from a rough copy ; these rough copies were written on the backs of manuscript or of proof-sheets, and were almost illegible, sometimes even to himself. He made a rule of keeping *all* letters that he received ; this was a habit which he learnt from his father, and which he said had been of great use to him.

He received many letters from foolish, unscrupulous people, and all of these received replies. He used to say that if he did not answer them, he had it on his conscience afterwards, and no doubt it was in great measure the courtesy with which he answered every one, which produced the universal and wide-spread sense of his kindness of nature, which was so evident on his death.

He was considerate to his correspondents in other and lesser things, for instance when dictating a letter to a foreigner he hardly ever failed to say to me, " You'd better try and write well, as it's to a foreigner." His letters were generally written on the assumption that they would be carelessly read ; thus, when he was dictating, he was careful to tell me to make an important clause begin with an obvious paragraph " to catch his eye," as he often said. How much he thought of the trouble he gave others by asking questions, will be well enough shown by his letters. It is difficult to say anything about the general

tone of his letters, they will speak for themselves. The unvarying courtesy of them is very striking. I had a proof of
this quality in the feeling with which Mr. Hacon, his solicitor,
regarded him. He had never seen my father, yet had a
sincere feeling of friendship for him, and spoke especially of
his letters as being such as a man seldom receives in the way
of business :—" Everything I did was right, and everything
was profusely thanked for."

  He had a printed form to be used in replying to troublesome
correspondents, but he hardly ever used it ; I suppose he never
found an occasion that seemed exactly suitable. I remember
an occasion on which it might have been used with advantage.
He received a letter from a stranger stating that the writer
had undertaken to uphold Evolution at a debating society,
and that being a busy young man, without time for reading,
he wished to have a sketch of my father's views. Even
this wonderful young man got a civil answer, though I
think he did not get much material for his speech. His rule
was to thank the donors of books, but not of pamphlets. He
sometimes expressed surprise that so few people thanked him
for his books which he gave away liberally ; the letters that
he did receive gave him much pleasure, because he habitually
formed so humble an estimate of the value of all his works,
that he was genuinely surprised at the interest which they
excited.

  In money and business matters he was remarkably careful
and exact. He kept accounts with great care, classifying
them, and balancing at the end of the year like a merchant.
I remember the quick way in which he would reach out for
his account-book to enter each cheque paid, as though he were
in a hurry to get it entered before he had forgotten it. His
father must have allowed him to believe that he would be
poorer than he really was, for some of the difficulty experienced in finding a house in the country must have arisen
from the modest sum he felt prepared to give. Yet he knew,

of course, that he would be in easy circumstances, for in his
'Recollections' he mentions this as one of the reasons for his
not having worked at medicine with so much zeal as he
would have done if he had been obliged to gain his living.
He had a pet economy in paper, but it was rather a hobby
than a real economy. All the blank sheets of letters received
were kept in a portfolio to be used in making notes ; it was
his respect for paper that made him write so much on the
backs of his old MS., and in this way, unfortunately, he destroyed large parts of the original MS. of his books. His
feeling about paper extended to waste paper, and he objected,
half in fun, to the careless custom of throwing a spill into the
fire after it had been used for lighting a candle.

My father was wonderfully liberal and generous to all his
children in the matter of money, and I have special cause to
remember his kindness when I think of the way in which he
paid some Cambridge debts of mine—making it almost seem
a virtue in me to have told him of them. In his later years
he had the kind and generous plan of dividing his surplus at
the year's end among his children.

He had a great respect for pure business capacity, and
often spoke with admiration of a relative who had doubled
his fortune. And of himself would often say in fun that
what he really *was* proud of was the money he had saved.
He also felt satisfaction in the money he made by his books.
His anxiety to save came in great measure from his fears
that his children would not have health enough to earn their
own livings, a foreboding which fairly haunted him for many
years. And I have a dim recollection of his saying, "Thank
God, you'll have bread and cheese," when I was so young that
I was rather inclined to take it literally.

When letters were finished, about three in the afternoon,
he rested in his bedroom, lying on the sofa and smoking
a cigarette, and listening to a novel or other book not
scientific. He only smoked when resting, whereas snuff

was a stimulant, and was taken during working hours. He took snuff for many years of his life, having learnt the habit at Edinburgh as a student. He had a nice silver snuff-box given him by Mrs. Wedgwood of Maer, which he valued much —but he rarely carried it, because it tempted him to take too many pinches. In one of his early letters he speaks of having given up snuff for a month, and describes himself as feeling " most lethargic, stupid and melancholy." Our former neighbour and clergyman, Mr. Brodie Innes, tells me that at one time my father made a resolve not to take snuff except away from home, " a most satisfactory arrangement for me," he adds, " as I kept a box in my study to which there was access from the garden without summoning servants, and I had more frequently, than might have been otherwise the case, the privilege of a few minutes' conversation with my dear friend." He generally took snuff from a jar on the hall table, because having to go this distance for a pinch was a slight check ; the clink of the lid of the snuff jar was a very familiar sound. Sometimes when he was in the drawing-room, it would occur to him that the study fire must be burning low, and when some of us offered to see after it, it would turn out that he also wished to get a pinch of snuff.

Smoking he only took to permanently of late years, though on his Pampas rides he learned to smoke with the Gauchos, and I have heard him speak of the great comfort of a cup of *maté* and a cigarette when he halted after a long ride and was unable to get food for some time.

The reading aloud often sent him to sleep, and he used to regret losing parts of a novel, for my mother went steadily on lest the cessation of the sound might wake him. He came down at four o'clock to dress for his walk, and he was so regular that one might be quite certain it was within a few minutes of four when his descending steps were heard.

From about half-past four to half-past five he worked ; then he came to the drawing-room, and was idle till it was time (about six) to go up for another rest with novel-reading and a cigarette.

Latterly he gave up late dinner, and had a simple tea at half-past seven (while we had dinner), with an egg or a small piece of meat. After dinner he never stayed in the room, and used to apologise by saying he was an old woman, who must be allowed to leave with the ladies. This was one of the many signs and results of his constant weakness and ill-health. Half an hour more or less conversation would make to him the difference of a sleepless night, and of the loss perhaps of half the next day's work.

After dinner he played backgammon with my mother, two games being played every night; for many years a score of the games which each won was kept, and in this score he took the greatest interest. He became extremely animated over these games, bitterly lamenting his bad luck and exploding with exaggerated mock-anger at my mother's good fortune.

After backgammon he read some scientific book to himself, either in the drawing-room, or, if much talking was going on, in the study.

In the evening, that is, after he had read as much as his strength would allow, and before the reading aloud began, he would often lie on the sofa and listen to my mother playing the piano. He had not a good ear, yet in spite of this he had a true love of fine music. He used to lament that his enjoyment of music had become dulled with age, yet within my recollection his love of a good tune was strong. I never heard him hum more than one tune, the Welsh song "Ar hyd y nos," which he went through correctly; he used also, I believe, to hum a little Otaheitan song. From his want of ear he was unable to recognize a tune when he heard it again, but he remained constant to what he liked, and would often say, when an old favourite was played, "That's a fine thing; what is it?" He liked especially parts of Beethoven's symphonies, and bits of Handel. He made a little list of all the pieces

which he especially liked among those which my mother
played—giving in a few words the impression that each one
made on him—but these notes are unfortunately lost. He
was sensitive to differences in style, and enjoyed the late Mrs.
Vernon Lushington's playing intensely, and in June 1881,
when Hans Richter paid a visit at Down, he was roused to
strong enthusiasm by his magnificent performance on the
piano. He much enjoyed good singing, and was moved
almost to tears by grand or pathetic songs. His niece Lady
Farrer's singing of Sullivan's " Will he come " was a never-
failing enjoyment to him. He was humble in the extreme
about his own taste, and correspondingly pleased when he
found that others agreed with him.

He became much tired in the evenings, especially of late
years, and left the drawing-room about ten, going to bed at
half-past ten. His nights were generally bad, and he often
lay awake or sat up in bed for hours, suffering much discomfort. He was troubled at night by the activity of his thoughts,
and would become exhausted by his mind working at some
problem which he would willingly have dismissed. At night,
too, anything which had vexed or troubled him in the day
would haunt him, and I think it was then that he suffered if
he had not answered some troublesome person's letter.

The regular readings, which I have mentioned, continued
for so many years, enabled him to get through a great
deal of the lighter kinds of literature. He was extremely
fond of novels, and I remember well the way in which
he would anticipate the pleasure of having a novel read
to him, as he lay down, or lighted his cigarette. He
took a vivid interest both in plot and characters, and
would on no account know before-hand, how a story
finished ; he considered looking at the end of a novel as a
feminine vice. He could not enjoy any story with a tragical
end, for this reason he did not keenly appreciate George
Eliot, though he often spoke warmly in praise of 'Silas

Marner.' Walter Scott, Miss Austen, and Mrs. Gaskell, were read and re-read till they could be read no more. He had two or three books in hand at the same time—a novel and perhaps a biography and a book of travels. He did not often read out-of-the-way or old standard books, but generally kept to the books of the day obtained from a circulating library.

I do not think that his literary tastes and opinions were on a level with the rest of his mind. He himself, though he was clear as to what he thought good, considered that in matters of literary taste, he was quite outside the pale, and often spoke of what those within it liked or disliked, as if they formed a class to which he had no claim to belong.

In all matters of art he was inclined to laugh at professed critics, and say that their opinions were formed by fashion. Thus in painting, he would say how in his day every one admired masters who are now neglected. His love ot pictures as a young man is almost a proof that he must have had an appreciation of a portrait as a work of art, not as a likeness. Yet he often talked laughingly of the small worth of portraits, and said that a photograph was worth any number of pictures, as if he were blind to the artistic quality in a painted portrait. But this was generally said in his attempts to persuade us to give up the idea of having his portrait painted, an operation very irksome to him.

This way of looking at himself as an ignoramus in all matters of art, was strengthened by the absence of pretence, which was part of his character. With regard to questions of taste, as well as to more serious things, he always had the courage of his opinions. I remember, however, an instance that sounds like a contradiction to this : when he was looking at the Turners in Mr. Ruskin's bedroom, he did not confess, as he did afterwards, that he could make out absolutely nothing of what Mr. Ruskin saw in them. But this little pretence was not for his own sake, but for the sake of courtesy to his host. He was pleased and amused when subsequently

Mr. Ruskin brought him some photographs of pictures (I think Vandyke portraits), and courteously seemed to value my father's opinion about them.

Much of his scientific reading was in German, and this was a great labour to him; in reading a book after him, I was often struck at seeing, from the pencil-marks made each day where he left off, how little he could read at a time. He used to call German the "Verdammte," pronounced as if in English. He was especially indignant with Germans, because he was convinced that they could write simply if they chose, and often praised Dr. F. Hildebrand for writing German which was as clear as French. He sometimes gave a German sentence to a friend, a patriotic German lady, and used to laugh at her if she did not translate it fluently. He himself learnt German simply by hammering away with a dictionary; he would say that his only way was to read a sentence a great many times over, and at last the meaning occurred to him. When he began German long ago, he boasted of the fact (as he used to tell) to Sir J. Hooker, who replied, "Ah, my dear fellow, that's nothing; I've begun it many times."

In spite of his want of grammar, he managed to get on wonderfully with German, and the sentences that he failed to make out were generally really difficult ones. He never attempted to speak German correctly, but pronounced the words as though they were English; and this made it not a little difficult to help him, when he read out a German sentence and asked for a translation. He certainly had a bad ear for vocal sounds, so that he found it impossible to perceive small differences in pronunciation.

His wide interest in branches of science that were not specially his own was remarkable. In the biological sciences his doctrines make themselves felt so widely that there was something interesting to him in most departments of it. He read a good deal of many quite special works, and large parts of text books, such as Huxley's 'Invertebrate Anatomy,' or such a book as Balfour's 'Embryology,' where the detail, at any rate, was not specially in his own line. And in the case of elaborate books of the monograph type, though he did not make a study of them, yet he felt the strongest admiration for them.

In the non-biological sciences he felt keen sympathy with work of which he could not really judge. For instance, he used to read nearly the whole of 'Nature,' though so much of it deals with mathematics and physics. I have often heard him say that he got a kind of satisfaction in reading articles which (according to himself) he could not understand. I wish I could reproduce the manner in which he would laugh at himself for it.

It was remarkable, too, how he kept up his interest in subjects at which he had formerly worked. This was strikingly the case with geology. In one of his letters to Mr. Judd he begs him to pay him a visit, saying that since Lyell's death he hardly ever gets a geological talk. His observations, made only a few years before his death, on the upright pebbles in the drift at Southampton, and discussed in a letter to Mr. Geikie, afford another instance. Again, in the letters to Dr. Dohrn, he shows how his interest in barnacles remained alive. I think it was all due to the vitality and persistence of his mind—a quality I have heard him speak of as if he felt that he was strongly gifted in that respect. Not that he used any such phrases as these about himself, but he would say that he had the power of keeping a subject or question more or less before him for a great many years. The extent to which he possessed this power appears when we consider the number of different problems which he solved, and the early period at which some of them began to occupy him.

It was a sure sign that he was not well when he was idle at any times other than his regular resting hours; for, as long as he remained moderately well, there was no break in the regularity of his life. Week-days and Sundays passed by alike, each with their stated intervals of work and rest. It is almost impossible, except for those who watched his daily life, to realise how essential to his well-being was the regular routine that I have sketched : and with what pain and difficulty anything beyond it was attempted. Any public appearance, even of the most modest kind, was an effort to him. In 1871 he went to the little village church for the wedding of his elder daughter, but he could hardly bear the fatigue of being present through the short service. The same may be said of the few other occasions on which he was present at similar ceremonies.

I remember him many years ago at a christening; a memory which has remained with me, because to us children it seemed an extraordinary and abnormal occurrence. I remember his look most distinctly at his brother Erasmus's funeral, as he stood in the scattering of snow, wrapped in a long black funeral cloak, with a grave look of sad reverie.

When, after an interval of many years, he again attended a meeting of the Linnean Society, it was felt to be, and was in fact, a serious undertaking ; one not to be determined on without much sinking of heart, and hardly to be carried into effect without paying a penalty of subsequent suffering. In the same way a breakfast-party at Sir James Paget's, with some of the distinguished visitors to the Medical Congress (1881), was to him a severe exertion.

The early morning was the only time at which he could make any effort of the kind, with comparative impunity. Thus it came about that the visits he paid to his scientific friends in London were by preference made as early as ten in the morning. For the same reason he started on his journeys by the earliest possible train, and used to arrive at the houses of relatives in London when they were beginning their day.

He kept an accurate journal of the days on which he worked and those on which his ill health prevented him from working, so that it would be possible to tell how many were idle days

in any given year. In this journal—a little yellow Letts's
Diary, which lay open on his mantel-piece, piled on the
diaries of previous years—he also entered the day on which
he started for a holiday and that of his return.

The most frequent holidays were visits of a week to
London, either to his brother's house (6 Queen Anne Street),
or to his daughter's (4 Bryanston Street). He was generally
persuaded by my mother to take these short holidays, when
it became clear from the frequency of "bad days," or from
the swimming of his head, that he was being overworked.
He went unwillingly, and tried to drive hard bargains, stipu-
lating, for instance, that he should come home in five days
instead of six. Even if he were leaving home for no more
than a week, the packing had to be begun early on the
previous day, and the chief part of it he would do himself.
The discomfort of a journey to him was, at least latterly,
chiefly in the anticipation, and in the miserable sinking feeling
from which he suffered immediately before the start ; even a
fairly long journey, such as that to Coniston, tired him wonderfully little, considering how much an invalid he was ; and he
certainly enjoyed it in an almost boyish way, and to a curious
extent.

Although, as he has said, some of his æsthetic tastes had
suffered a gradual decay, his love of scenery remained fresh
and strong. Every walk at Coniston was a fresh delight, and
he was never tired of praising the beauty of the broken hilly
country at the head of the lake.

One of the happy memories of this time [1879] is that of a
delightful visit to Grasmere : "The perfect day," my sister
writes, "and my father's vivid enjoyment and flow of spirits,
form a picture in my mind that I like to think of. He
could hardly sit still in the carriage for turning round and
getting up to admire the view from each fresh point, and even
in returning he was full of the beauty of Rydal Water, though

he would not allow that Grasmere at all equalled his beloved Coniston."

Besides these longer holidays, there were shorter visits to various relatives—to his brother-in-law's house, close to Leith Hill, and to his son near Southampton. He always particularly enjoyed rambling over rough open country, such as the commons near Leith Hill and Southampton, the heath-covered wastes of Ashdown Forest, or the delightful "Rough" near the house of his friend Sir Thomas Farrer. He never was quite idle even on these holidays, and found things to observe. At Hartfield he watched Drosera catching insects, &c.; at Torquay he observed the fertilisation of an orchid (*Spiranthes*), and also made out the relations of the sexes in Thyme.

He was always rejoiced to get home after his holidays; he used greatly to enjoy the welcome he got from his dog Polly, who would get wild with excitement, panting, squeaking, rushing round the room, and jumping on and off the chairs; and he used to stoop down, pressing her face to his, letting her lick him, and speaking to her with a peculiarly tender, caressing voice.

My father had the power of giving to these summer holidays a charm which was strongly felt by all his family. The pressure of his work at home kept him at the utmost stretch of his powers of endurance, and when released from it, he entered on a holiday with a youthfulness of enjoyment that made his companionship delightful; we felt that we saw more of him in a week's holiday than in a month at home.

Some of these absences from home, however, had a depressing effect on him; when he had been previously much overworked it seemed as though the absence of the customary strain allowed him to fall into a peculiar condition of miserable health.

Besides the holidays which I have mentioned, there were his

visits to water-cure establishments. In 1849, when very ill, suffering from constant sickness, he was urged by a friend to try the water-cure, and at last agreed to go to Dr. Gully's establishment at Malvern. His letters to Mr. Fox show how much good the treatment did him; he seems to have thought that he had found a cure for his troubles, but, like all other remedies, it had only a transient effect on him. However, he found it, at first, so good for him, that when he came home he built himself a douche-bath, and the butler learnt to be his bathman.

He paid many visits to Moor Park, Dr. Lane's water-cure establishment in Surrey, not far from Aldershot. These visits were pleasant ones, and he always looked back to them with pleasure. Dr. Lane has given his recollections of my father in Dr. Richardson's 'Lecture on Charles Darwin,' October 22, 1882, from which I quote:—

"In a public institution like mine, he was surrounded, of course, by multifarious types of character, by persons of both sexes, mostly very different from himself—commonplace people, in short, as the majority are everywhere, but like to him at least in this, that they were fellow-creatures and fellow-patients. And never was any one more genial, more considerate, more friendly, more altogether charming than he universally was." . . . . He "never aimed, as too often happens with good talkers, at monopolising the conversation. It was his pleasure rather to give and take, and he was as good a listener as a speaker. He never preached nor prosed, but his talk, whether grave or gay (and it was each by turns), was full of life and salt—racy, bright, and animated."

Some idea of his relation to his family and his friends may be gathered from what has gone before; it would be impossible to attempt a complete account of these relationships, but a slightly fuller outline may not be out of place. Of his

K 2

married life I cannot speak, save in the briefest manner. In his relationship towards my mother, his tender and sympathetic nature was shown in its most beautiful aspect. In her presence he found his happiness, and through her, his life,—which might have been overshadowed by gloom,—became one of content and quiet gladness.

The 'Expression of the Emotions' shows how closely he watched his children; it was characteristic of him that (as I have heard him tell), although he was so anxious to observe accurately the expression of a crying child, his sympathy with the grief spoiled his observation. His note-book, in which are recorded sayings of his young children, shows his pleasure in them. He seemed to retain a sort of regretful memory of the childhoods which had faded away, and thus he wrote in his 'Recollections':—" When you were very young it was my delight to play with you all, and I think with a sigh that such days can never return."

I may quote, as showing the tenderness of his nature, some sentences from an account of his little daughter Annie, written a few days after her death :—

"Our poor child, Annie, was born in Gower Street, on March 2, 1841, and expired at Malvern at mid-day on the 23rd of April, 1851.

"I write these few pages, as I think in after years, if we live, the impressions now put down will recall more vividly her chief characteristics. From whatever point I look back at her, the main feature in her disposition which at once rises before me, is her buoyant joyousness, tempered by two other characteristics, namely, her sensitiveness, which might easily have been overlooked by a stranger, and her strong affection. Her joyousness and animal spirits radiated from her whole countenance, and rendered every movement elastic and full of life and vigour. It was delightful and cheerful to behold her. Her dear face now rises before me, as she used sometimes to come running downstairs with a stolen pinch of snuff for me her whole form radiant with the pleasure of giving pleasure. Even when playing with her cousins, when her joyousness almost passed into boisterousness, a single glance of my eye, not of displeasure (for I thank God I hardly ever cast one on her), but of want of sympathy, would for some minutes alter her whole countenance.

" The other point in her character, which made her joyousness and spirits so delightful, was her strong affection, which was of a most clinging, fondling nature. When quite a baby, this showed itself in never being easy without touching her mother, when in bed with her ; and quite lately she would, when poorly, fondle for any length of time one of her mother's arms. When very unwell, her mother lying down beside her, seemed to soothe her in a manner quite different from what it would have done to any of our other children. So, again, she would at almost any time spend half an hour in arranging my hair, 'making it,' as she called it, 'beautiful,' or in smoothing, the poor dear darling, my collar or cuffs—in short, in fondling me.

" Besides her joyousness thus tempered, she was in her manners remarkably cordial, frank, open, straightforward, natural, and without any shade of reserve. Her whole mind was pure and transparent. One felt one knew her thoroughly and could trust her. I always thought, that come what might, we should have had in our old age, at least one loving soul, which nothing could have changed. All her movements were vigorous, active, and usually graceful. When going round the Sand-walk with me, although I walked fast, yet she often used to go before, pirouetting in the most elegant way, her dear face bright all the time with the sweetest smiles. Occasionally she had a pretty coquettish manner towards me, the memory of which is charming. She often used exaggerated language, and when I quizzed her by exaggerating what she had said, how clearly can

I now see the little toss of the head, and exclamation of, 'Oh, papa, what a shame of you!' In the last short illness, her conduct in simple truth was angelic. She never once complained; never became fretful; was ever considerate of others, and was thankful in the most gentle, pathetic manner for everything done for her. When so exhausted that she could hardly speak, she praised everything that was given her, and said some tea 'was beautifully good.' When I gave her some water, she said, 'I quite thank you;' and these, I believe, were the last precious words ever addressed by her dear lips to me.

"We have lost the joy of the household, and the solace of our old age. She must have known how we loved her. Oh, that she could now know how deeply, how tenderly, we do still and shall ever love her dear joyous face! Blessings on her!

"April 30, 1851."

We his children all took especial pleasure in the games he played at with us, but I do not think he romped much with us; I suppose his health prevented any rough play. He used sometimes to tell us stories, which were considered specially delightful, partly on account of their rarity.

The way he brought us up is shown by a little story about my brother Leonard, which my father was fond of telling. He came into the drawing-room and found Leonard dancing about on the sofa, which was forbidden, for the sake of the springs, and said, "Oh, Lenny, Lenny, that's against all rules," and received for answer, "Then I think you'd better go out of the room." I do not believe he ever spoke an angry word to any of his children in his life; but I am certain that it never entered our heads to disobey him. I well remember one occasion when my father reproved me for a piece of carelessness; and I can still recall the feeling of depression which came over me, and the care which he took to disperse it by speaking to me soon afterwards with especial kindness. He kept up his delightful, affectionate manner towards us all his life. I sometimes wonder that he could do so, with such an undemonstrative race as we are ; but I hope he knew how much we delighted in his loving words and manner. How often, when a man, I have wished when my father was behind my chair, that he would pass his hand over my hair, as he used to do when I was a boy. He allowed his grown-up children to laugh with and at him, and was generally speaking on terms of perfect equality with us.

He was always full of interest about each one's plans or successes. We used to laugh at him, and say he would not believe in his sons, because, for instance, he would be a little doubtful about their taking some bit of work for which he did not feel sure that they had knowledge enough. On the other hand, he was only too much inclined to take a favourable view of our work. When I thought he had set too high a value on anything that I had done, he used to be indignant and inclined to explode in mock anger. His doubts were part of his humility concerning what was in any way connected with himself; his too favourable view of our work was due to his sympathetic nature, which made him lenient to every one.

He kept up towards his children his delightful manner of expressing his thanks ; and I never wrote a letter, or read a page aloud to him, without receiving a few kind words of recognition. His love and goodness towards his little grandson Bernard were great ; and he often spoke of the pleasure it was to him to see "his little face opposite to him" at luncheon. He and Bernard used to compare their tastes ; e.g., in liking brown sugar better than white, &c. ; the result being, "We always agree, don't we ? "

My sister writes :—

"My first remembrances of my father are of the delights of his playing with us. He was passionately attached to his

own children, although he was not an indiscriminate child-
lover. To all of us he was the most delightful play-fellow,
and the most perfect sympathiser. Indeed it is impossible
adequately to describe how delightful a relation his was to his
family, whether as children or in their later life.

"It is a proof of the terms on which we were, and also of
how much he was valued as a play-fellow, that one of his sons
when about four years old tried to bribe him with sixpence to
come and play in working hours. We all knew the sacred-
ness of working time, but that any one should resist sixpence
seemed an impossibility.

"He must have been the most patient and delightful of
nurses. I remember the haven of peace and comfort it
seemed to me when I was unwell, to be tucked up on the
study sofa, idly considering the old geological map hung on
the wall. This must have been in his working hours, for I
always picture him sitting in the horse-hair arm-chair by the
corner of the fire.

"Another mark of his unbounded patience was the way in
which we were suffered to make raids into the study when we
had an absolute need of sticking-plaster, string, pins, scissors,
stamps, foot-rule, or hammer. These and other such neces-
saries were always to be found in the study, and it was the
only place where this was a certainty. We used to feel it
wrong to go in during work-time ; still, when the necessity was
great, we did so. I remember his patient look when he said
once, 'Don't you think you could not come in again, I have
been interrupted very often.' We used to dread going in for
sticking-plaster, because he disliked to see that we had cut
ourselves, both for our sakes and on account of his acute
sensitiveness to the sight of blood. I well remember lurking
about the passage till he was safe away, and then stealing
in for the plaster.

"Life seems to me, as I look back upon it, to have been very
regular in those early days, and except relations (and a few

intimate friends), I do not think any one came to the house. After lessons, we were always free to go where we would, and that was chiefly in the drawing-room and about the garden, so that we were very much with both my father and mother. We used to think it most delightful when he told us any stories about the *Beagle*, or about early Shrewsbury days— little bits about school-life and his boyish tastes. Sometimes too he read aloud to his children such books as Scott's novels, and I remember a few little lectures on the steam-engine.

"I was more or less ill during the five years between my thirteenth and eighteenth years, and for a long time (years it seems to me) he used to play a couple of games of back-gammon with me every afternoon. He played them with the greatest spirit, and I remember we used at one time to keep account of the games, and as this record came out in favour of him, we kept a list of the doublets thrown by each, as I was convinced that he threw better than myself.

"His patience and sympathy were boundless during this weary illness, and sometimes when most miserable I felt his sympathy to be almost too keen. When at my worst, we went to my aunt's house at Hartfield, in Sussex, and as soon as we had made the move safely he went on to Moor Park for a fortnight's water-cure. I can recall now how on his return I could hardly bear to have him in the room, the expression of tender sympathy and emotion in his face was too agitating, coming fresh upon me after his little absence.

"He cared for all our pursuits and interests, and lived our lives with us in a way that very few fathers do. But I am certain that none of us felt that this intimacy interfered the least with our respect or obedience. Whatever he said was absolute truth and law to us. He always put his whole mind into answering any of our questions. One trifling instance makes me feel how he cared for what we cared for. He had no special taste for cats, though he admired the pretty ways of a kitten. But yet he knew and remembered the individualities of my many cats, and would talk about the habits and characters of the more remarkable ones years after they had died.

"Another characteristic of his treatment of his children was his respect for their liberty, and for their personality. Even as quite a girl, I remember rejoicing in this sense of freedom. Our father and mother would not even wish to know what we were doing or thinking unless we wished to tell. He always made us feel that we were each of us creatures whose opinions and thoughts were valuable to him, so that whatever there was best in us came out in the sunshine of his presence.

"I do not think his exaggerated sense of our good qualities, intellectual or moral, made us conceited, as might perhaps have been expected, but rather more humble and grateful to him. The reason being no doubt that the influence of his character, of his sincerity and greatness of nature, had a much deeper and more lasting effect than any small exaltation which his praises or admiration may have caused to our vanity."

As head of a household he was much loved and respected ; he always spoke to servants with politeness, using the expression, "would you be so good," in asking for anything. He was hardly ever angry with his servants ; it shows how seldom this occurred, that when, as a small boy, I overheard a servant being scolded, and my father speaking angrily, it impressed me as an appalling circumstance, and I remember running up stairs out of a general sense of awe. He did not trouble himself about the management of the garden, cows, &c. He considered the horses so little his concern, that he used to ask doubtfully whether he might have a horse and cart to send to Keston for Drosera, or to the Westerham nurseries for plants, or the like.

As a host my father had a peculiar charm : the presence of visitors excited him, and made him appear to his best advantage. At Shrewsbury, he used to say, it was his father's wish that the guests should be attended to constantly, and in one of the letters to Fox he speaks of the impossibility of writing a letter while the house was full of company. I think he always felt uneasy at not doing more for the entertainment of his guests, but the result was successful; and, to make up for any loss, there was the gain that the guests felt perfectly free to do as they liked. The most usual visitors were those who stayed from Saturday till Monday ; those who remained longer were generally relatives, and were considered to be rather more my mother's affair than his.

Besides these visitors, there were foreigners and other strangers, who came down for luncheon and went away in the afternoon. He used conscientiously to represent to them the enormous distance of Down from London, and the labour it would be to come there, unconsciously taking for granted that they would find the journey as toilsome as he did himself. If, however, they were not deterred, he used to arrange their journeys for them, telling them when to come, and practically when to go. It was pleasant to see the way in which he shook hands with. a guest who was being welcomed for the first time ; his hand used to shoot out in a way that gave one the feeling that it was hastening to meet the guest's hands. With old friends his hand came down with a hearty swing into the other hand in a way I always had satisfaction in seeing. His good-bye was chiefly characterised by the pleasant way in which he thanked his guests, as he stood at the door, for having come to see him.

These luncheons were very successful entertainments, there was no drag or flagging about them, my father was bright and excited throughout the whole visit. Professor De Candolle has described a visit to Down, in his admirable and sympathetic sketch of my father.* He speaks of his manner

* ' Darwin considéré au point de vue des causes de son succès.'—Geneva, 1882.

as resembling that of a "savant" of Oxford or Cambridge. This does not strike me as quite a good comparison; in his ease and naturalness there was more of the manner of some soldiers; a manner arising from total absence of pretence or affectation. It was this absence of pose, and the natural and simple way in which he began talking to his guests, so as to get them on their own lines, which made him so charming a host to a stranger. His happy choice of matter for talk seemed to flow out of his sympathetic nature, and humble, vivid interest in other people's work.

To some, I think, he caused actual pain by his modesty; I have seen the late Francis Balfour quite discomposed by having knowledge ascribed to himself on a point about which my father claimed to be utterly ignorant.

It is difficult to seize on the characteristics of my father's conversation.

He had more dread than have most people of repeating his stories, and continually said, "You must have heard me tell," or "I dare say I've told you." One peculiarity he had, which gave a curious effect to his conversation. The first few words of a sentence would often remind him of some exception to, or some reason against, what he was going to say; and this again brought up some other point, so that the sentence would become a system of parenthesis within parenthesis, and it was often impossible to understand the drift of what he was saying until he came to the end of his sentence. He used to say of himself that he was not quick enough to hold an argument with any one, and I think this was true. Unless it was a subject on which he was just then at work, he could not get the train of argument into working order quickly enough. This is shown even in his letters; thus, in the case of two letters to Prof. Semper about the effect of isolation, he did not recall the series of facts he wanted until some days after the first letter had been sent off.

When puzzled in talking, he had a peculiar stammer on the

first word of a sentence. I only recall this occurring with words beginning with w; possibly he had a special difficulty with this letter, for I have heard him say that as a boy he could not pronounce w, and that sixpence was offered him if he could say "white wine," which he pronounced "rite rine." Possibly he may have inherited this tendency from Erasmus Darwin, who stammered.*

He sometimes combined his metaphors in a curious way, using such a phrase as "holding on like life,"—a mixture of "holding on for his life," and "holding on like grim death." It came from his eager way of putting emphasis into what he was saying. This sometimes gave an air of exaggeration where it was not intended; but it gave, too, a noble air of strong and generous conviction; as, for instance, when he gave his evidence before the Royal Commission on vivisection and came out with his words about cruelty, "It deserves detestation and abhorrence." When he felt strongly about any similar question, he could hardly trust himself to speak, as he then easily became angry, a thing which he disliked excessively. He was conscious that his anger had a tendency to multiply itself in the utterance, and for this reason dreaded (for example) having to scold a servant.

It was a great proof of the modesty of his style of talking, that, when, for instance, a number of visitors came over from Sir John Lubbock's for a Sunday afternoon call, he never seemed to be preaching or lecturing, although he had so much of the talk to himself. He was particularly charming when "chaffing" any one, and in high spirits over it. His manner at such times was light-hearted and boyish, and his refinement of nature came out most strongly. So, when he was talking to a lady who pleased and amused him, the combina-

* My father related a Johnsonian answer of Erasmus Darwin's: "Don't you find it very inconvenient stammering, Dr. Darwin?" "No, sir, because I have time to think before I speak, and don't ask impertinent questions."

tion of raillery and deference in his manner was delightful
to see.

When my father had several guests he managed them well,
getting a talk with each, or bringing two or three together
round his chair. In these conversations there was always
a good deal of fun, and, speaking generally, there was either
a humorous turn in his talk, or a sunny geniality which
served instead. Perhaps my recollection of a pervading element of humour is the more vivid, because the best talks were
with Mr. Huxley, in whom there is the aptness which is akin
to humour, even when humour itself is not there. My father
enjoyed Mr. Huxley's humour exceedingly, and would often
say, " What splendid fun Huxley is ! "   I think he probably
had more scientific argument (of the nature of a fight) with
Lyell and Sir Joseph Hooker.

He used to say that it grieved him to find that for the
friends of his later life he had not the warm affection of
his youth. Certainly in his early letters from Cambridge
he gives proofs of very strong friendship for Herbert and
Fox ; but no one except himself would have said that his
affection for his friends was not, throughout life, of the
warmest possible kind. In serving a friend he would not
spare himself, and precious time and strength were willingly
given. He undoubtedly had, to an unusual degree, the power
of attaching his friends to him. He had many warm friend-
ships, but to Sir Joseph Hooker he was bound by ties of
affection stronger than we often see among men. He wrote
in his ' Recollections,' " I have known hardly any man more
lovable than Hooker."

His relationship to the village people was a pleasant one ;
he treated them, one and all, with courtesy, when he came in
contact with them, and took an interest in all relating to
their welfare. Some time after he came to live at Down he
helped to found a Friendly Club, and served as treasurer for
thirty years. He took much trouble about the club, keep-

ing its accounts with minute and scrupulous exactness, and taking pleasure in its prosperous condition. Every Whit-Monday the club used to march round with band and banner, and paraded on the lawn in front of the house. There he met them, and explained to them their financial position in a little speech seasoned with a few well-worn jokes. He was often unwell enough to make even this little ceremony an exertion, but I think he never failed to meet them.

He was also treasurer of the Coal Club, which gave him some work, and he acted for some years as a County Magistrate.

With regard to my father's interest in the affairs of the village, Mr. Brodie Innes has been so good as to give me his recollections :—

"On my becoming Vicar of Down in 1846, we became friends, and so continued till his death. His conduct towards me and my family was one of unvarying kindness, and we repaid it by warm affection.

"In all parish matters he was an active assistant; in matters connected with the schools, charities, and other business, his liberal contribution was ever ready, and in the differences which at times occurred in that, as in other parishes, I was always sure of his support. He held that where there was really no important objection, his assistance should be given to the clergyman, who ought to know the circumstances best, and was chiefly responsible."

His intercourse with strangers was marked with scrupulous and rather formal politeness, but in fact he had few opportunities of meeting strangers.

Dr. Lane has described * how, on the rare occasion of my father attending a lecture (Dr. Sanderson's) at the Royal Institution, "the whole assembly . . . rose to their feet to welcome him," while he seemed "scarcely conscious that such an outburst of applause could possibly be intended for himself."

* Lecture by Dr. B. W. Richardson, in St. George's Hall, Oct. 22, 1882.

The quiet life he led at Down made him feel confused in a large society; for instance, at the Royal Society's *soirées* he felt oppressed by the numbers. The feeling that he ought to know people, and the difficulty he had in remembering faces in his latter years, also added to his discomfort on such occasions. He did not realise that he would be recognised from his photographs, and I remember his being uneasy at being obviously recognised by a stranger at the Crystal Palace Aquarium.

I must say something of his manner of working: one characteristic of it was his respect for time; he never forgot how precious it was. This was shown, for instance, in the way in which he tried to curtail his holidays; also, and more clearly, with respect to shorter periods. He would often say, that saving the minutes was the way to get work done; he showed this love of saving the minutes in the difference he felt between a quarter of an hour and ten minutes' work; he never wasted a few spare minutes from thinking that it was not worth while to set to work. I was often struck by his way of working up to the very limit of his strength, so that he suddenly stopped in dictating, with the words, " I believe I mustn't do any more." The same eager desire not to lose time was seen in his quick movements when at work. I particularly remember noticing this when he was making an experiment on the roots of beans, which required some care in manipulation; fastening the little bits of card upon the roots was done carefully and necessarily slowly, but the intermediate movements were all quick; taking a fresh bean, seeing that the root was healthy, impaling it on a pin, fixing it on a cork, and seeing that it was vertical, &c.; all these processes were performed with a kind of restrained eagerness. He always gave one the impression of working with pleasure, and not with any drag. I have an image, too, of him as he recorded the result of some experiment, looking eagerly at each root, &c., and then writing with equal eagerness. I remember the quick movement of his head up and down as he looked from the object to the notes.

He saved a great deal of time through not having to do things twice. Although he would patiently go on repeating experiments where there was any good to be gained, he could not endure having to repeat an experiment which ought, if complete care had been taken, to have succeeded the first time—and this gave him a continual anxiety that the experiment should not be wasted ; he felt the experiment to be sacred, however slight a one it was. He wished to learn as much as possible from an experiment, so that he did not confine himself to observing the single point to which the experiment was directed, and his power of seeing a number of other things was wonderful. I do not think he cared for preliminary or rough observations intended to serve as guides and to be repeated. Any experiment done was to be of some use, and in this connection I remember how strongly he urged the necessity of keeping the notes of experiments which failed, and to this rule he always adhered.

In the literary part of his work he had the same horror of losing time, and the same zeal in what he was doing at the moment, and this made him careful not to be obliged unnecessarily to read anything a second time.

His natural tendency was to use simple methods and few instruments. The use of the compound microscope has much increased since his youth, and this at the expense of the simple one. It strikes us nowadays as extraordinary that he should have had no compound microscope when he went his *Beagle* voyage ; but in this he followed the advice of Robt. Brown, who was an authority in such matters. He always had a great liking for the simple microscope, and maintained that nowadays it was too much neglected, and that one ought always to see as much as possible with the simple before taking to the compound microscope. In one of his letters he speaks on this point, and remarks that he always suspects the work of a man who never uses the simple microscope.

His dissecting table was a thick board, let into a window of the study; it was lower than an ordinary table, so that he could not have worked at it standing; but this, from wishing to save his strength, he would not have done in any case. He sat at his dissecting-table on a curious low stool which had belonged to his father, with a seat revolving on a vertical spindle, and mounted on large castors, so that he could turn easily from side to side. His ordinary tools, &c., were lying about on the table, but besides these a number of odds and ends were kept in a round table full of radiating drawers, and turning on a vertical axis, which stood close by his left side, as he sat at his microscope-table. The drawers were labelled, "best tools," "rough tools," "specimens," "preparations for specimens," &c. The most marked peculiarity of the contents of these drawers was the care with which little scraps and almost useless things were preserved; he held the well-known belief, that if you threw a thing away you were sure to want it directly—and so things accumulated.

If any one had looked at his tools, &c., lying on the table, he would have been struck by an air of simpleness, make-shift, and oddness.

At his right hand were shelves, with a number of other odds and ends, glasses, saucers, tin biscuit boxes for germinating seeds, zinc labels, saucers full of sand, &c., &c. Considering how tidy and methodical he was in essential things, it is curious that he bore with so many make-shifts: for instance, instead of having a box made of a desired shape, and stained black inside, he would hunt up something like what he wanted and get it darkened inside with shoe-blacking; he did not care to have glass covers made for tumblers in which he germinated seeds, but used broken bits of irregular shape, with perhaps a narrow angle sticking uselessly out on one side. But so much of his experimenting was of a simple

kind, that he had no need for any elaboration, and I think
his habit in this respect was in great measure due to his
desire to husband his strength, and not waste it on inessential
things.

His way of marking objects may here be mentioned. If
he had a number of things to distinguish, such as leaves,
flowers, &c., he tied threads of different colours round them.
In particular he used this method when he had only two
classes of objects to distinguish; thus in the case of crossed
and self-fertilised flowers, one set would be marked with
black and one with white thread, tied round the stalk of
the flower. I remember well the look of two sets of capsules, gathered and waiting to be weighed, counted, &c.,
with pieces of black and of white thread to distinguish the
trays in which they lay. When he had to compare two
sets of seedlings, sowed in the same pot, he separated them
by a partition of zinc-plate; and the zinc-label, which gave
the necessary details about the experiment, was always
placed on a certain side, so that it became instinctive with
him to know without reading the label which were the
" crossed " and which the " self-fertilised."

His love of each particular experiment, and his eager zeal
not to lose the fruit of it, came out markedly in these crossing experiments—in the elaborate care he took not to make
any confusion in putting capsules into wrong trays, &c., &c.
I can recall his appearance as he counted seeds under the
simple microscope with an alertness not usually characterising
such mechanical work as counting. I think he personified
each seed as a small demon trying to elude him by getting
into the wrong heap, or jumping away altogether; and this
gave to the work the excitement of a game. He had great
faith in instruments, and I do not think it naturally occurred
to him to doubt the accuracy of a scale or measuring
glass, &c. He was astonished when we found that one of his
micrometers differed from the other. He did not require any

great accuracy in most of his measurements, and had not good scales; he had an old three-foot rule, which was the common property of the household, and was constantly being borrowed, because it was the only one which was certain to be in its place—unless, indeed, the last borrower had forgotten to put it back. For measuring the height of plants, he had a seven-foot deal rod, graduated by the village carpenter Latterly he took to using paper scales graduated to milli-meters. For small objects he used a pair of compasses and an ivory protractor. It was characteristic of him that he took scrupulous pains in making measurements with his somewhat rough scales. A trifling example of his faith in authority is that he took his "inch in terms of millimeters" from an old book, in which it turned out to be inaccurately given. He had a chemical balance which dated from the days when he worked at chemistry with his brother Erasmus. Measure-ments of capacity were made with an apothecary's measuring glass: I remember well its rough look and bad graduation. With this, too, I remember the great care he took in getting the fluid-line on to the graduation. I do not mean by this account of his instruments that any of his experiments suffered from want of accuracy in measurement, I give them as examples of his simple methods and faith in others—faith at least in instrument-makers, whose whole trade was a mystery to him.

A few of his mental characteristics, bearing especially on his mode of working, occur to me. There was one quality of mind which seemed to be of special and extreme advantage in leading him to make discoveries. It was the power of never letting exceptions pass unnoticed. Everybody notices a fact as an exception when it is striking or frequent, but he had a special instinct for arresting an exception. A point appa-rently slight and unconnected with his present work is passed over by many a man almost unconsciously with some half-considered explanation, which is in fact no explanation. It was just these things that he seized on to make a start from. In a certain sense there is nothing special in this procedure, many discoveries being made by means of it. I only mention it because, as I watched him at work, the value of this power to an experimenter was so strongly impressed upon me.

Another quality which was shown in his experimental work, was his power of sticking to a subject ; he used almost to apologise for his patience, saying that he could not bear to be beaten, as if this were rather a sign of weakness on his part. He often quoted the saying, " It's dogged as does it ;" and I think doggedness expresses his frame of mind almost better than perseverance. Perseverance seems hardly to express his almost fierce desire to force the truth to reveal itself. He often said that it was important that a man should know the right point at which to give up an inquiry. And I think it was his tendency to pass this point that inclined him to apologise for his perseverance, and gave the air of doggedness to his work.

He often said that no one could be a good observer unless he was an active theoriser. This brings me back to what I said about his instinct for arresting exceptions : it was as though he were charged with theorising power ready to flow into any channel on the slightest disturbance, so that no fact, however small, could avoid releasing a stream of theory, and thus the fact became magnified into importance. In this way it naturally happened that many untenable theories occurred to him ; but fortunately his richness of imagination was equalled by his power of judging and condemning the thoughts that occurred to him. He was just to his theories, and did not condemn them unheard ; and so it happened that he was willing to test what would seem to most people not at all worth testing. These rather wild trials he called " fool's experiments," and enjoyed extremely. As an example I may mention that finding the cotyledons of Biophytum to be highly sensitive to vibrations of the table, he fancied that they might perceive the vibrations of sound, and therefore made me play my bassoon close to a plant.*

The love of experiment was very strong in him, and I can remember the way he would say, "I shan't be easy till I have tried it," as if an outside force were driving him. He enjoyed experimenting much more than work which only entailed reasoning, and when he was engaged on one of his books which required argument and the marshalling of facts, he felt experimental work to be a rest or holiday. Thus, while working upon the 'Variations of Animals and Plants,' in 1860–61, he made out the fertilisation of Orchids, and thought himself idle for giving so much time to them. It is interesting to think that so important a piece of research should have been undertaken and largely worked out as a pastime in place of more serious work. The letters to Hooker of this period contain expressions such as, "God forgive me for being so idle; I am quite sillily interested in the work." The intense pleasure he took in understanding the adaptations for fertilisation is strongly shown in these letters. He speaks in one of his letters of his intention of working at Drosera as a rest from the 'Descent of Man.' He has described in his 'Recollections' the strong satisfaction he felt in solving the problem of heterostylism. And I have heard him mention that the Geology of South America gave him almost more pleasure than anything else. It was perhaps this delight in work requiring keen observation that made him value praise given to his observing powers almost more than appreciation of his other qualities.

For books he had no respect, but merely considered them as tools to be worked with. Thus he did not bind them, and even when a paper book fell to pieces from use, as happened to Müller's 'Befruchtung,' he preserved it from complete dissolution by putting a metal clip over its back. In the same

---

* This is not so much an example of superabundant theorising from a small cause, but only of his wish to test the most improbable ideas.

way he would cut a heavy book in half, to make it more convenient to hold. He used to boast that he had made Lyell
publish the second edition of one of his books in two volumes,
instead of in one, by telling him how he had been obliged to
cut it in half. Pamphlets were often treated even more
severely than books, for he would tear out, for the sake of
saving room, all the pages except the one that interested him.
The consequence of all this was, that his library was not
ornamental, but was striking from being so evidently a working
collection of books.

He was methodical in his manner of reading books and
pamphlets bearing on his own work. He had one shelf on
which were piled up the books he had not yet read, and
another to which they were transferred after having been
read, and before being catalogued. He would often groan
over his unread books, because there were so many which he
knew he should never read. Many a book was at once transferred to the other heap, either marked with a cypher at the
end, to show that it contained no marked passages, or in-
scribed, perhaps, "not read," or "only skimmed." The books
accumulated in the "read" heap until the shelves overflowed,
and then, with much lamenting, a day was given up to the
cataloguing. He disliked this work, and as the necessity
of undertaking the work became imperative, would often
say, in a voice of despair, "We really must do these books
soon."

In each book, as he read it, he marked passages bearing on
his work. In reading a book or pamphlet, &c., he made
pencil-lines at the side of the page, often adding short
remarks, and at the end made a list of the pages marked.
When it was to be catalogued and put away, the marked
pages were looked at, and so a rough abstract of the book
was made. This abstract would perhaps be written under
three or four headings on different sheets, the facts being
sorted out and added to the previously collected facts in

different subjects. He had other sets of abstracts arranged, not according to subject, but according to periodical. When collecting facts on a large scale, in earlier years, he used to read through, and make abstracts, in this way, of whole series of periodicals.

In some of his early letters he speaks of filling several note-books with facts for his book on species; but it was certainly early that he adopted his plan of using portfolios, as described in the 'Recollections.' * My father and M. de Candolle were mutually pleased to discover that they had adopted the same plan of classifying facts. De Candolle describes the method in his ' Phytologie,' and in his sketch of my father mentions the satisfaction he felt in seeing it in action at Down.

Besides these portfolios, of which there are some dozens full of notes, there are large bundles of MS. marked "used" and put away. He felt the value of his notes, and had a horror of their destruction by fire. I remember, when some alarm of fire had happened, his begging me to be especially careful, adding very earnestly, that the rest of his life would be miserable if his notes and books were to be destroyed.

He shows the same feeling in writing about the loss of a manuscript, the purport of his words being, " I have a copy, or the loss would have killed me." In writing a book he would spend much time and labour in making a skeleton or plan of the whole, and in enlarging and sub-classing each heading, as described in his 'Recollections.' I think this careful arrangement of the plan was not at all essential to the building up of his argument, but for its presentment, and for the arrangement of his facts. In his 'Life of Erasmus Darwin,' as it was first printed in slips, the growth of the book from a skeleton was plainly visible. The arrangement

---

* The racks in which the port-folios were placed are shown in the illustration at the head of the chapter, in the recess at the right-hand side of the fire-place.

was altered afterwards, because it was too formal and cate-
gorical, and seemed to give the character of his grandfather
rather by means of a list of qualities than as a complete
picture.

It was only within the last few years that he adopted a plan
of writing which he was convinced suited him best, and which
is described in the 'Recollections'; namely, writing a rough
copy straight off without the slightest attention to style. It
was characteristic of him that he felt unable to write with
sufficient want of care if he used his best paper, and thus it
was that he wrote on the backs of old proofs or manuscript.
The rough copy was then reconsidered, and a fair copy was
made. For this purpose he had foolscap paper ruled at wide
intervals, the lines being needed to prevent him writing so
closely that correction became difficult. The fair copy was
then corrected, and was recopied before being sent to the
printers. The copying was done by Mr. E. Norman, who
began this work many years ago when village schoolmaster at
Down. My father became so used to Mr. Norman's hand-
writing, that he could not correct manuscript, even when
clearly written out by one of his children, until it had been
recopied by Mr. Norman. The MS., on returning from Mr.
Norman, was once more corrected, and then sent off to the
printers. Then came the work of revising and correcting the
proofs, which my father found especially wearisome.

It was at this stage that he first seriously considered the
style of what he had written. When this was going on he
usually started some other piece of work as a relief. The
correction of slips consisted in fact of two processes, for the
corrections were first written in pencil, and then re-considered
and written in ink.

When the book was passing through the "slip" stage he
was glad to have corrections and suggestions from others.
Thus my mother looked over the proofs of the 'Origin.' In
some of the later works my sister, Mrs. Litchfield, did much

of the correction. After my sister's marriage perhaps most of the work fell to my share.

My sister, Mrs. Litchfield, writes :—

"This work was very interesting in itself, and it was inexpressibly exhilarating to work for him. He was always so ready to be convinced that any suggested alteration was an improvement, and so full of gratitude for the trouble taken. I do not think that he ever used to forget to tell me what improvement he thought I had made, and he used almost to excuse himself if he did not agree with any correction. I think I felt the singular modesty and graciousness of his nature through thus working for him in a way I never should otherwise have done.

"He did not write with ease, and was apt to invert his sentences both in writing and speaking, putting the qualifying clause before it was clear what it was to qualify. He corrected a great deal, and was eager to express himself as well as he possibly could."

Perhaps the commonest corrections needed were of obscurities due to the omission of a necessary link in the reasoning, something which he had evidently omitted through familiarity with the subject. Not that there was any fault in the sequence of the thoughts, but that from familiarity with his argument he did not notice when the words failed to reproduce his thought. He also frequently put too much matter into one sentence, so that it had to be cut up into two.

On the whole, I think the pains which my father took over the literary part of the work was very remarkable. He often laughed or grumbled at himself for the difficulty which he found in writing English, saying, for instance, that if a bad arrangement of a sentence was possible, he should be sure to adopt it. He once got much amusement and satisfaction out of the difficulty which one of the family found in writing a short circular. He had the pleasure of correcting and laughing at obscurities, involved sentences, and other defects, and thus took his revenge for all the criticism he had himself to bear with. He used to quote with astonishment Miss Martineau's advice to young authors, to write straight off and send the MS. to the printer without correction. But in some cases he acted in a somewhat similar manner. When a sentence got hopelessly involved, he would ask himself, "now what *do* you want to say?" and his answer written down, would often disentangle the confusion.

His style has been much praised; on the other hand, at least one good judge has remarked to me that it is not a good style. It is, above all things, direct and clear; and it is characteristic of himself in its simplicity, bordering on naïveté, and in its absence of pretence. He had the strongest disbelief in the common idea that a classical scholar must write good English; indeed, he thought that the contrary was the case. In writing, he sometimes showed the same tendency to strong expressions as he did in conversation. Thus in the 'Origin,' p. 440, there is a description of a larval cirripede, "with six pairs of beautifully constructed natatory legs, a pair of magnificent compound eyes, and extremely complex antennæ." We used to laugh at him for this sentence, which we compared to an advertisement. This tendency to give himself up to the enthusiastic turn of his thought, without fear of being ludicrous, appears elsewhere in his writings.

His courteous and conciliatory tone towards his reader is remarkable, and it must be partly this quality which revealed his personal sweetness of character to so many who had never seen him. I have always felt it to be a curious fact, that he who has altered the face of Biological Science, and is in this respect the chief of the moderns, should have written and worked in so essentially a non-modern spirit and manner. In reading his books one is reminded of the older naturalists rather than of the modern school of writers. He was a Naturalist in the old sense of

the word, that is, a man who works at many branches of
science, not merely a specialist in one. Thus it is, that,
though he founded whole new divisions of special subjects
—such as the fertilisation of flowers, insectivorous plants,
dimorphism, &c.—yet even in treating these very subjects
he does not strike the reader as a specialist. The reader
feels like a friend who is being talked to by a courteous
gentleman, not like a pupil being lectured by a professor.
The tone of such a book as the 'Origin' is charming, and
almost pathetic ; it is the tone of a man who, convinced of the
truth of his own views, hardly expects to convince others ; it
is just the reverse of the style of a fanatic, who wants to
force people to believe. The reader is never scorned for any
amount of doubt which he may be imagined to feel, and his
scepticism is treated with patient respect. A sceptical reader,
or perhaps even an unreasonable reader, seems to have been
generally present to his thoughts. It was in consequence of
this feeling, perhaps, that he took much trouble over points
which he imagined would strike the reader, or save him
trouble, and so tempt him to read.

    For the same reason he took much interest in the illus-
trations of his books, and I think rated rather too highly
their value. The illustrations for his earlier books were
drawn by professional artists. This was the case in 'Animals
and Plants,' the 'Descent of Man,' and the 'Expression of
the Emotions.' On the other hand, 'Climbing Plants,'
'Insectivorous Plants,' the 'Movements of Plants,' and
'Forms of Flowers,' were, to a large extent, illustrated by
some of his children—my brother George having drawn by
far the most. It was delightful to draw for him, as he was
enthusiastic in his praise of very moderate performances. I
remember well his charming manner of receiving the drawings
of one of his daughters-in-law, and how he would finish his
words of praise by saying, "Tell A——, Michael Angelo is
nothing to it." Though he praised so generously, he always

looked closely at the drawing, and easily detected mistakes or carelessness.

He had a horror of being lengthy, and seems to have been really much annoyed and distressed when he found how the 'Variations of Animals and Plants' was growing under his hands. I remember his cordially agreeing with 'Tristram Shandy's' words, "Let no man say, 'Come, I'll write a duodecimo.'"

His consideration for other authors was as marked a characteristic as his tone towards his reader. He speaks of all other authors as persons deserving of respect. In cases where, as in the case of ——'s experiments on Drosera, he thought lightly of the author, he speaks of him in such a way that no one would suspect it. In other cases he treats the confused writings of ignorant persons as though the fault lay with himself for not appreciating or understanding them. Besides this general tone of respect, he had a pleasant way of expressing his opinion on the value of a quoted work, or his obligation for a piece of private information.

His respectful feeling was not only morally beautiful, but was I think of practical use in making him ready to consider the ideas and observations of all manner of people. He used almost to apologise for this, and would say that he was at first inclined to rate everything too highly.

It was a great merit in his mind that, in spite of having so strong a respectful feeling towards what he read, he had the keenest of instincts as to whether a man was trustworthy or not. He seemed to form a very definite opinion as to the accuracy of the men whose books he read; and made use of this judgment in his choice of facts for use in argument or as illustrations. I gained the impression that he felt this power of judging of a man's trustworthiness to be of much value.

He had a keen feeling of the sense of honour that ought to reign among authors, and had a horror of any kind of laxness in quoting. He had a contempt for the love of honour and glory, and in his letters often blames himself for the pleasure he took in the success of his books, as though he were departing from his ideal—a love of truth and carelessness about fame. Often, when writing to Sir J. Hooker what he calls a boasting letter, he laughs at himself for his conceit and want of modesty. There is a wonderfully interesting letter which he wrote to my mother bequeathing to her, in case of his death, the care of publishing the manuscript of his first essay on evolution. This letter seems to me full of the intense desire that his theory should succeed as a contribution to knowledge, and apart from any desire for personal fame. He certainly had the healthy desire for success which a man of strong feelings ought to have. But at the time of the publication of the 'Origin' it is evident that he was overwhelmingly satisfied with the adherence of such men as Lyell, Hooker, Huxley, and Asa Gray, and did not dream of or desire any such wide and general fame as he attained to.

Connected with his contempt for the undue love of fame, was an equally strong dislike of all questions of priority. The letters to Lyell, at the time of the ' Origin,' show the anger he felt with himself for not being able to repress a feeling of disappointment at what he thought was Mr. Wallace's forestalling of all his years of work. His sense of literary honour comes out strongly in these letters ; and his feeling about priority is again shown in the admiration expressed in his ' Recollections' of Mr. Wallace's self-annihilation.

His feeling about reclamations, including answers to attacks and all kinds of discussions, was strong. It is simply expressed in a letter to Falconer (1863), "If I ever felt angry towards you, for whom I have a sincere friendship, I should begin to suspect that I was a little mad. I was very sorry about your reclamation, as I think it is in every case a mistake and should be left to others. Whether I should so act myself under provocation is a different question." It was a feeling partly dictated by instinctive delicacy, and partly by a strong sense of the waste of time, energy, and temper thus caused. He said that he owed his determination not to get into discussions * to the advice of Lyell,—advice which he transmitted to those among his friends who were given to paper warfare.

If the character of my father's working life is to be understood, the conditions of ill-health, under which he worked, must be constantly borne in mind. He bore his illness with such uncomplaining patience, that even his children can hardly, I believe, realise the extent of his habitual suffering. In their case the difficulty is heightened by the fact that, from the days of their earliest recollections, they saw him in constant ill-health,—and saw him, in spite of it, full of pleasure in what pleased them. Thus, in later life, their perception of what he endured had to be disentangled from the impression produced in childhood by constant genial kindness under conditions of unrecognised difficulty. No one indeed, except my mother, knows the full amount of suffering he endured, or the full amount of his wonderful patience. For all the latter years of his life she never left him for a night; and her days were so planned that all his resting hours might be shared with her. She shielded him from every avoidable annoyance, and omitted nothing that might save him trouble, or prevent him becoming overtired, or that might alleviate the many discomforts of his ill-health. I hesitate to speak thus freely of a thing so sacred as the life-long devotion which prompted all this constant

---

* He departed from his rule in his "Note on the Habits of the Pampas Woodpecker, *Colaptes campestris*," 'Proc. Zool. Soc.,' 1870, p. 705 : also in a letter published in the 'Athenæum' (1863, p. 554), in which case he afterwards regretted that he had not remained silent. His replies to criticisms, in the later editions of the 'Origin,' can hardly be classed as infractions of his rule.

and tender care. But it is, I repeat, a principal feature of his life, that for nearly forty years he never knew one day of the health of ordinary men, and that thus his life was one long struggle against the weariness and strain of sickness. And this cannot be told without speaking of the one condition which enabled him to bear the strain and fight out the struggle to the end.

LETTERS.

———◆———

THE earliest letters to which I have access are those written by my father when an undergraduate at Cambridge.

The history of his life, as told in his correspondence, must therefore begin with this period.

CHAPTER IV.

CAMBRIDGE LIFE.

[MY father's Cambridge life comprises the time between the Lent Term, 1828, when he came up as a Freshman, and the end of the May Term, 1831, when he took his degree and left the University.

It appears from the College books, that my father "admissus est pensionarius minor sub Magistro Shaw" on Oct. 15, 1827. He did not come into residence till the Lent Term, 1828, so that, although he passed his examination in due season, he was unable to take his degree at the usual time,—the beginning of the Lent Term, 1831. In such a case a man usually took his degree before Ash-Wednesday, when he was called "Baccalaureus ad Diem Cinerum," and ranked with the B.A.'s of the year. My father's name, however, occurs in the list of Bachelors "ad Baptistam," or those admitted between Ash-Wednesday and St. John Baptist's Day (June 24th); * he therefore took rank among the Bachelors of 1832.

He "kept" for a term or two in lodgings, over Bacon the tobacconist's ; not, however, over the shop in the Market Place, now so well known to Cambridge men, but in Sidney Street. For the rest of his time he had pleasant rooms on the south side of the first court of Christ's.†

What determined the choice of this college for his brother

---

* "On Tuesday last Charles Darwin, of Christ's College, was admitted B.A."—*Cambridge Chronicle*, Friday, April 29, 1831.

† The rooms are on the first floor, on the west side of the middle staircase. A medallion (given by my brother) has recently been let into the wall of the sitting-room.

Erasmus and himself I have no means of knowing. Erasmus
the elder, their grandfather, had been at St. John's, and this
college might have been reasonably selected for them, being
connected with Shrewsbury School. But the life of an undergraduate at St. John's seems, in those days, to have been a
troubled one, if I may judge from the fact that a relative
of mine migrated thence to Christ's to escape the harassing
discipline of the place. A story told by Mr. Herbert * illustrates the same state of things :—

"In the beginning of the October Term of 1830, an incident
occurred which was attended with somewhat disagreeable,
though ludicrous consequences to myself. Darwin asked me
to take a long walk with him in the Fens, to search for some
natural objects he was desirous of having. After a very long,
fatiguing day's work, we dined together, late in the evening,
at his rooms in Christ's College ; and as soon as our dinner
was over we threw ourselves into easy chairs and fell sound
asleep. I was the first to awake, about three in the morning,
when, having looked at my watch, and knowing the strict
rule of St. John's, which required men *in statu pupillari* to
come into college before midnight, I rushed homeward at the
utmost speed, in fear of the consequences, but hoping that
the Dean would accept the excuse as sufficient when I told
him the real facts. He, however, was inexorable, and refused
to receive my explanations, or any evidence I could bring ;
and although during my undergraduateship I had never been
reported for coming late into College, now, when I was a hard-
working B.A., and had five or six pupils, he sentenced me to
confinement to the College walls for the rest of the term.
Darwin's indignation knew no bounds, and the stupid injustice
and tyranny of the Dean raised not only a perfect ferment
among my friends, but was the subject of expostulation from
some of the leading members of the University."

My father seems to have found no difficulty in living at

* See footnote, p. 49.

peace with all men in and out of office at Lady Margaret's
other foundation. The impression of a contemporary of my
father's is that Christ's in their day was a pleasant, fairly quiet
college, with some tendency towards "horsiness"; many of
the men made a custom of going to Newmarket during the
races, though betting was not a regular practice. In this they
were by no means discouraged by the Senior Tutor, Mr.
Shaw, who was himself generally to be seen on the Heath on
these occasions. There was a somewhat high proportion of
Fellow-Commoners,—eight or nine, to sixty or seventy Pen-
sioners, and this would indicate that it was not an unpleasant
college for men with money to spend and with no great love
of strict discipline.

The way in which the service was conducted in chapel
shows that the Dean, at least, was not over zealous. I have
heard my father tell how at evening chapel the Dean used to
read alternate verses of the Psalms, without making even a
pretence of waiting for the congregation to take their share.
And when the Lesson was a lengthy one, he would rise and
go on with the Canticles after the scholar had read fifteen or
twenty verses.

It is curious that my father often spoke of his Cambridge
life as if it had been so much time wasted, forgetting that,
although the set studies of the place were barren enough for
him, he yet gained in the highest degree the best advantages
of a University life—the contact with men and an opportunity
for his mind to grow vigorously. It is true that he valued
at its highest the advantages which he gained from associating with Professor Henslow and some others, but he seemed
to consider this as a chance outcome of his life at Cambridge,
not an advantage for which *Alma Mater* could claim any
credit. One of my father's Cambridge friends was the late
Mr. J. M. Herbert, County Court Judge for South Wales,
from whom I was fortunate enough to obtain some notes
which help us to gain an idea of how my father impressed

his contemporaries.   Mr. Herbert writes : " I think it was in
the spring of 1828 that I first met Darwin, either at my
cousin Whitley's rooms in St. John's, or at the rooms of some
other of his old Shrewsbury schoolfellows, with many of
whom I was on terms of great intimacy.   But it certainly was
in the summer of that year that our acquaintance ripened
into intimacy, when we happened to be together at Barmouth,
 or the Long Vacation, reading with private tutors,—he with
Betterton of St. John's, his Classical and Mathematical Tutor,
and I with Yate of St. John's."

The intercourse between them practically ceased in 1831,
when my father said good-bye to Herbert at Cambridge, on
starting on his *Beagle* voyage.   I once met Mr. Herbert,
then almost an old man, and I was much struck by the
evident warmth and freshness of the affection with which he
remembered my father. The notes from which I quote end with
this warm-hearted eulogium : " It would be idle for me to speak
of his vast intellectual powers . . . but I cannot end this cur-
sory and rambling sketch without testifying, and I doubt not
all his surviving college friends would concur with me, that
he was the most genial, warm-hearted, generous, and affectionate of friends ; that his sympathies were with all that was
good and true ; and that he had a cordial hatred for everything false, or vile, or cruel, or mean, or dishonourable.   He
was not only great, but pre-eminently good, and just, and
loveable."

Two anecdotes told by Mr. Herbert show that my father's
feeling for suffering, whether of man or beast, was as strong
in him as a young man as it was in later years : " Before he
left Cambridge he told me that he had made up his mind not
to shoot any more ; that he had had two days' shooting at his
friend's, Mr. Owen of Woodhouse ; and that on the second
day, when going over some of the ground they had beaten
on the day before, he picked up a bird not quite dead,
but lingering from a shot it had received on the pre-

vious day ; and that it had made and left such a painful
impression on his mind, that he could not reconcile it to his
conscience to continue to derive pleasure from a sport which
inflicted such cruel suffering."

To realise the strength of the feeling that led to this resolve, we must remember how passionate was his love of
sport. We must recall the boy shooting his first snipe,* and
trembling with excitement so that he could hardly reload his
gun. Or think of such a sentence as, " Upon my soul, it is
only about a fortnight to the ' First,' then if there is a bliss
on earth that is it." †

Another anecdote told by Mr. Herbert illustrates again his
tenderness of heart :—

" When at Barmouth, he and I went to an exhibition of
' learned dogs.' In the middle of the entertainment one of
the dogs failed in performing the trick his master told him
to do. On the man reproving him, the dog put on a most
piteous expression, as if in fear of the whip. Darwin seeing
it, asked me to leave with him, saying, ' Come along, I can't
stand this any longer ; how those poor dogs must have been
licked.' "

It is curious that the same feeling recurred to my father
more than fifty years afterwards, on seeing some performing dogs at the Westminster Aquarium ; on this occasion
he was reassured by the manager telling him that the dogs
were taught more by reward than by punishment. Mr. Herbert
goes on :—" It stirred one's inmost depth of feeling to hear
him descant upon, and groan over, the horrors of the slave
trade, or the cruelties to which the suffering Poles were subjected to at Warsaw. . . . These, and other like proofs have
left on my mind the conviction that a more humane or
tender-hearted man never lived."

* ' Recollections,' p. 34.
† Letter from C. Darwin to W. D. Fox.

His old college friends agree in speaking with affectionate warmth of his pleasant, genial temper as a young man. From what they have been able to tell me, I gain the impression of a young man overflowing with animal spirits—leading a varied healthy life—not over-industrious in the set studies of the place, but full of other pursuits, which were followed with a rejoicing enthusiasm. Entomology, riding, shooting in the fens, suppers and card-playing, music at King's Chapel, engravings at the Fitzwilliam Museum, walks with Professor Henslow—all combined to fill up a happy life. He seems to have infected others with his enthusiasm. Mr. Herbert relates how, during the same Barmouth summer, he was pressed into the service of " the science "—as my father called collecting beetles. They took their daily walks together among the hills behind Barmouth, or boated in the Mawddach estuary, or sailed to Sarn Badrig to land there at low water, or went fly-fishing in the Cors-y-gedol lakes. " On these occasions Darwin entomologised most industriously, picking up creatures as he walked along, and bagging everything which seemed worthy of being pursued, or of further examination. And very soon he armed me with a bottle of alcohol, in which I had to drop any beetle which struck me as not of a common kind. I performed this duty with some diligence in my constitutional walks ; but alas ! my powers of discrimination seldom enabled me to secure a prize—the usual result, on his examining the contents of my bottle, being an exclamation, ' Well, old Cherbury ' * (the nickname he gave me, and by which he usually addressed me), 'none of these will do.'" Again, the Rev. T. Butler, who was one of the Barmouth reading-party in 1828, says : " He inoculated me with a taste for Botany which has stuck by me all my life."

Archdeacon Watkins, another old college friend of my father's, remembers him unearthing beetles in the willows

* No doubt in allusion to the title of Lord Herbert of Cherbury.

between Cambridge and Grantchester, and speaks of a certain beetle the remembrance of whose name is " Crux major." * How enthusiastically must my father have exulted over this beetle to have impressed its name on a companion so that he remembers it after half a century ! Archdeacon Watkins goes on : " I do not forget the long and very interesting conversations that we had about Brazilian scenery and tropical vegetation of all sorts. Nor do I forget the way and the vehemence with which he rubbed his chin when he got excited on such subjects, and discoursed eloquently of lianas, orchids, &c."

He became intimate with Henslow, the Professor of Botany, and through him with some other older members of the University. "But," Mr. Herbert writes, " he always kept up the closest connection with the friends of his own standing ; and at our frequent social gatherings—at breakfast, wine or supper parties—he was ever one of the most cheerful, the most popular, and the most welcome."

· My father formed one of a club for dining once a week, called the Gourmet † Club, the members, besides himself and Mr. Herbert (from whom I quote), being Whitley of St. John's, now Honorary Canon of Durham ; ‡ Heaviside of Sidney, now Canon of Norwich ; Lovett Cameron of Trinity, now vicar of Shoreham ; Blane of Trinity, who held a high post during the Crimean war ; H. Lowe § (now Sherbrooke) of Trinity Hall ; and Watkins of Emmanuel, now Archdeacon of York. The origin of the club's name seems already to have become involved in obscurity. Mr. Herbert says that it was chosen in derision of another "set of men who called themselves by a long Greek name signifying 'fond of dainties,' but who falsified their claim to such a designation by their weekly practice of dining at some roadside inn, six miles from

* *Panagæus crux-major.*
† Mr. Herbert mentions the name as ' The Glutton Club.'
‡ Formerly Reader in Natural Philosophy at Durham University.
§ Brother of Lord Sherbrooke.

Cambridge, on mutton chops or beans and bacon." Another old member of the club tells me that the name arose because the members were given to making experiments on "birds and beasts which were before unknown to human palate." He says that hawk and bittern were tried, and that their zeal broke down over an old brown owl, "which was indescribable." At any rate, the meetings seemed to have been successful, and to have ended with "a game of mild vingt-et-un."

Mr. Herbert gives an amusing account of the musical examinations described by my father in his 'Recollections.' Mr. Herbert speaks strongly of his love of music, and adds, "What gave him the greatest delight was some grand symphony or overture of Mozart's or Beethoven's, with their full harmonies." On one occasion Herbert remembers "accompanying him to the afternoon service at King's, when we heard a very beautiful anthem. At the end of one of the parts, which was exceedingly impressive, he turned round to me and said, with a deep sigh, 'How's your backbone?'" He often spoke of a feeling of coldness or shivering in his back on hearing beautiful music.

Besides a love of music, he had certainly at this time a love of fine literature; and Mr. Cameron tells me that he used to read Shakespeare to my father in his rooms at Christ's, who took much pleasure in it. He also speaks of his "great liking for first-class line engravings, especially those of Raphael Morghen and Müller; and he spent hours in the Fitzwilliam Museum in looking over the prints in that collection."

My father's letters to Fox show how sorely oppressed he felt by the reading for an examination : "I am reading very hard, and have spirits for nothing. I actually have not stuck a beetle this term." His despair over mathematics must have been profound, when he expressed a hope that Fox's silence is due to "your being ten fathoms deep in the Mathe-

matics ; and if you are, God help you, for so am I, only with
this difference, I stick fast in the mud at the bottom, and
there I shall remain." Mr. Herbert says : " He had, I imagine,
no natural turn for mathematics, and he gave up his mathematical reading before he had mastered the first part of
algebra, having had a special quarrel with Surds and the
Binomial Theorem."

We get some evidence from his letters to Fox of my father's
intention of going into the Church. " I am glad," he writes,*
"to hear that you are reading divinity. I should like to know
what books you are reading, and your opinions about them ;
you need not be afraid of preaching to me prematurely."
Mr. Herbert's sketch shows how doubts arose in my father's
mind as to the possibility of his taking Orders. He writes,
"We had an earnest conversation about going into Holy
Orders ; and I remember his asking me, with reference to
the question put by the Bishop in the ordination service,
'Do you trust that you are inwardly moved by the Holy
Spirit, &c.,' whether I could answer in the affirmative, and on
my saying I could not, he said, 'Neither can I, and therefore
I cannot take orders.'" This conversation appears to have
taken place in 1829, and if so, the doubts here expressed
must have been quieted, for in May 1830, he speaks of having
some thoughts of reading divinity with Henslow.

The greater number of the following letters are addressed
by my father to his cousin, William Darwin Fox. Mr. Fox's
relationship to my father is shown in the pedigree given in
Chapter I. The degree of kinship appears to have remained
a problem to my father, as he signs himself in one letter
" $\frac{\text{cousin}}{n^2}$." Their friendship was, in fact, due to their being
undergraduates together. My father's letters show clearly
enough how genuine the friendship was. In after years, distance, large families, and ill-health on both sides, checked the

* March 18, 1829.

intercourse ; but a warm feeling of friendship remained. The correspondence was never quite dropped and continued till Mr. Fox's death in 1880. Mr. Fox took orders, and worked as a country clergyman until forced by ill-health to leave his living in Delamere Forest. His love of natural history remained strong, and he became a skilled fancier of many kinds of birds, &c. The index to 'Animals and Plants,' and my father's later correspondence, show how much help he received from his old College friend.]

C. Darwin to J. M. Herbert.

Saturday Evening
[September 14, 1828].*

MY DEAR OLD CHERBURY,

I am about to fulfil my promise of writing to you, but I am sorry to add there is a very selfish motive at the bottom. I am going to ask you a great favour, and you cannot imagine how much you will oblige me by procuring some more specimens of some insects which I dare say I can describe. In the first place, I must inform you that I have taken some of the rarest of the British Insects, and their being found near Barmouth, is quite unknown to the Entomological world : I think I shall write and inform some of the crack entomologists.

But now for business. *Several* more specimens, if you can procure them without much trouble, of the following insects :—The violet-black coloured beetle, found on Craig Storm, † under stones, also a large smooth black one very like it ; a bluish metallic-coloured dung-beetle, which is *very* common on the hill-sides ; also, if you *would* be so very kind as to cross the ferry, and you will find a great number under

---

* The postmark being Derby seems to show that the letter was written from his cousin, W. D. Fox's house, Osmaston, near Derby.

† The top of the hill immediately behind Barmouth was called Craig-Storm, a hybrid Cambro-English word.

the stones on the waste land of a long, smooth, jet-black beetle (a great many of these); also, in the same situation, a very small pinkish insect, with black spots, with a curved thorax projecting beyond the head; also, upon the marshy land over the ferry, near the sea, under old sea-weed, stones, &c., you will find a small yellowish transparent beetle, with two or four blackish marks on the back. Under these stones there are two sorts, one much darker than the other; the lighter-coloured is that which I want. These last two insects are *excessively rare*, and you will really *extremely* oblige me by taking all this trouble pretty soon. Remember me most kindly to Butler, tell him of my success, and I dare say both of you will easily recognise these insects. I hope his caterpillars go on well. I think many of the Chrysalises are well worth keeping. I really am quite ashamed [of] so long a letter all about my own concerns; but do return good for evil, and send me a long account of all your proceedings.

In the first week I killed seventy-five head of game—a very contemptible number—but there are very few birds. I killed, however, a brace of black game. Since then I have been staying at the Fox's, near Derby; it is a very pleasant house, and the music meeting went off very well. I want to hear how Yates likes his gun, and what use he has made of it.

If the bottle is not large you can buy another for me, and when you pass through Shrewsbury you can leave these treasures, and I hope, if you possibly can, you will stay a day or two with me, as I hope I need not say how glad I shall be to see you again. Fox remarked what deuced good-natured fellows your friends at Barmouth must be; and if I did not know that you and Butler were so, I would not think of giving you so much trouble.

Believe me, my dear Herbert,

Yours, most sincerely,

CHARLES DARWIN.

Remember me to all friends.

[In the following January we find him looking forward with pleasure to the beginning of another year of his Cambridge life : he writes to Fox—

"I waited till to-day for the chance of a letter, but I will wait no longer.   I must most sincerely and cordially congratulate you on having finished all your labours.   I think your place a *very good* one considering by how much you have beaten many men who had the start of you in reading. I do so wish I were now in Cambridge (a very selfish wish, however, as I was not with you in all your troubles and misery), to join in all the glory and happiness, which dangers gone by can give.   How we would talk, walk, and entomologise !   Sappho should be the best of bitches, and Dash, of dogs : then should be 'peace on earth, good will to men,'— which, by the way, I always think the most perfect description of happiness that words can give."]

C. Darwin to W. D. Fox.

Cambridge, Thursday [February 26, 1829].

MY DEAR FOX,

When I arrived here on Tuesday I found to my great grief and surprise, a letter on my table which I had written to you about a fortnight ago, the stupid porter never took the trouble of getting the letter forwarded.   I suppose you have been abusing me for a most ungrateful wretch ; but I am sure you will pity me now, as nothing is so vexatious as having written a letter in vain.

Last Thursday I left Shrewsbury for London, and stayed there till Tuesday, on which I came down here by the 'Times.' The first two days I spent entirely with Mr. Hope,* and did little else but talk about and look at insects ; his collection is most magnificent, and he himself is the most generous of entomologists ; he has given me about 160 new species, and

* Founder of the Chair of Zoology at Oxford.

actually often wanted to give me the rarest insects of which he had only two specimens. He made many civil speeches, and hoped you will call on him some time with me, whenever we should happen to be in London. He greatly compliments our exertions in Entomology, and says we have taken a wonderfully great number of good insects. On Sunday I spent the day with Holland, who lent me a horse to ride in the Park with.

On Monday evening I drank tea with Stephens;* his cabinet is more magnificent than the most zealous entomologist could dream of; he appears to be a very good-humoured pleasant little man. Whilst in town I went to the Royal Institution, Linnean Society, and Zoological Gardens, and many other places where naturalists are gregarious. If you had been with me, I think London would be a very delightful place; as things were, it was much pleasanter than I could have supposed such a dreary wilderness of houses to be.

I shot whilst in Shrewsbury a Dundiver (female Goosander, as I suppose you know). Shaw has stuffed it, and when I have an opportunity I will send it to Osmaston. There have been shot also five Waxen Chatterers, three of which Shaw has for sale; would you like to purchase a specimen? I have not yet thanked you for your last very long and agreeable letter. It would have been still more agreeable had it contained the joyful intelligence that you were coming up here; my two solitary breakfasts have already made me aware how very very much I shall miss you.

*   *   *   *   *

Believe me,

My dear old Fox,

Most sincerely yours,

C. DARWIN.

* J. F. Stephens, author of 'A Manual of British Coleoptera,' 1839, and other works.

[Later on in the Lent term he writes to Fox :—

"I am leading a quiet everyday sort of a life ; a little of Gibbon's History in the morning, and a good deal of *Van John* in the evening ; this, with an occasional ride with Simcox and constitutional with Whitley, makes up the regular routine of my days.   I see a good deal both of Herbert and Whitley, and the more I see of them increases every day the respect I have for their excellent understandings and dispositions. They have been giving some very gay parties, nearly sixty men there both evenings."]

C. Darwin to W. D. Fox.

Christ's College [Cambridge], April 1 [1829].

MY DEAR FOX,

In your letter to Holden you are pleased to observe "that of all the blackguards you ever met with I am the greatest." Upon this observation I shall make no remarks, excepting that I must give you all due credit for acting on it most rigidly.   And now I should like to know in what one particular are you less of a blackguard than I am ?   You idle old wretch, why have you not answered my last letter, which I am sure I forwarded to Clifton nearly three weeks ago ?   If I was not really very anxious to hear what you are doing, I should have allowed you to remain till you thought it worth while to treat me like a gentleman.   And now having vented my spleen in scolding you, and having told you, what you must know, how very much and how anxiously I want to hear how you and your family are getting on at Clifton, the purport of this letter is finished.   If you did but know how often I think of you, and how often I regret your absence, I am sure I should have heard from you long enough ago.

I find Cambridge rather stupid, and as I know scarcely any one that walks, and this joined with my lips not being quite so well, has reduced me to a sort of hybernation. . . . I have caught Mr. Harbour letting —— have the first pick of the beetles ; accordingly we have made our final adieus, my part in the affecting scene consisted in telling him he was a d—d rascal, and signifying I should kick him down the stairs if ever he appeared in my rooms again. It seemed altogether mightily to surprise the young gentleman. I have no news to tell you ; indeed, when a correspondence has been broken off like ours has been, it is difficult to make the first start again. Last night there was a terrible fire at Linton, eleven miles from Cambridge. Seeing the reflection so plainly in the sky, Hall, Woodyeare, Turner, and myself thought we would ride and see it. We set out at half-past nine, and rode like incarnate devils there, and did not return till two in the morning. Altogether it was a most awful sight. I cannot conclude without telling you, that of all the blackguards I ever met with, you are the greatest and the best.

C. DARWIN.

C. Darwin to W. D. Fox.

[Cambridge, Thursday, April 23, 1829.]

MY DEAR FOX,

I have delayed answering your last letter for these few days, as I thought that under such melancholy circumstances my writing to you would be probably only giving you trouble. This morning I received a letter from Catherine informing me of that event,* which, indeed, from your letter, I had hardly dared to hope would have happened otherwise. I feel most sincerely and deeply for you and all your family ; but at the same time, as far as any one can, by his own good principles and religion, be supported under such a misfortune, you, I am assured, will know where to look for such support. And after so pure and holy a comfort as the Bible affords, I am equally assured how useless the sympathy of all friends

---

* The death of Fox's sister, Mrs. Bristowe.

must appear, although it be as heartfelt and sincere, as I hope you believe me capable of feeling. At such a time of deep distress I will say nothing more, excepting that I trust your father and Mrs. Fox bear this blow as well as, under such circumstances, can be hoped for.

I am afraid it will be a long time, my dear Fox, before we meet ; till then, believe me at all times,

Yours most affectionately,

CHARLES DARWIN.

C. Darwin to W. D. Fox.

Shrewsbury, Friday [July 4, 1829].

MY DEAR FOX,

I should have written to you before only that whilst our expedition lasted I was too much engaged, and the conclusion was so unfortunate, that I was too unhappy to write to you till this week's quiet at home. The thoughts of Woodhouse next week has at last given me courage to relate my unfortunate case.

I started from this place about a fortnight ago to take an entomological trip with Mr. Hope through all North Wales ; and Barmouth was our first destination. The two first days I went on pretty well, taking several good insects ; but for the rest of that week my lips became suddenly so bad,* and I myself not very well, that I was unable to leave the room, and on the Monday I retreated with grief and sorrow back again to Shrewsbury. The first two days I took some good insects. . . . But the days that I was unable to go out, Mr. Hope did wonders . . . . and to-day I have received another parcel of insects from him, such Colymbetes, such Carabi, and such magnificent Elaters (two species of the bright scarlet sort). I am sure you will properly sympathise with my unfortunate situation : I am determined I will go over the

---

* Probably with eczema, from which he often suffered.

same ground that he does before autumn comes, and if working
hard will procure insects I will bring home a glorious stock.

*　　　　*　　　　*　　　　*　　　　*

My dear Fox,

Yours most sincerely,

CHAS. DARWIN.

C. Darwin to W. D. Fox.

Shrewsbury, July 18, 1829.

I am going to Maer next week in order to entomologise,
and shall stay there a week, and for the rest of this summer
I intend to lead a perfectly idle and wandering life. . . .
You see I am much in the same state that you are, with this
difference, you make good resolutions and never keep them ;
I never make them, so cannot keep them ; it is all very well
writing in this manner, but I must read for my Little-go.
Graham smiled and bowed so very civilly, when he told me
that he was one of the six appointed to make the examination
stricter, and that they were determined this would make it a
very different thing from any previous examination, that from
all this I am sure it will be the very devil to pay amongst all
idle men and entomologists.　Erasmus, we expect home in a
few weeks' time : he intends passing next winter in Paris.　Be
sure you order the two lists of insects published by Stephens,
one printed on both sides, and the other only on one ; you
will find them very useful in many points of view.

Dear old Fox, yours,

C. DARWIN.

C. Darwin to W. D. Fox.

Christ's College, Thursday [October 16, 1829].

MY DEAR FOX,

I am afraid you will be very angry with me for not
having written during the Music Meeting, but really I was

worked so hard that I had no time ; I arrived here on Monday and found my rooms in dreadful confusion, as they have been taking up the floor, and you may suppose that I have had plenty to do for these two days. The Music Meeting * was the most glorious thing I ever experienced ; and as for Malibran, words cannot praise her enough, she is quite the most charming person I ever saw. We had extracts out of several of the best operas, acted in character, and you cannot imagine how very superior it made the concerts to any I ever heard before. J. de Begnis † acted ' Il Fanatico ' in character ; being dressed up an extraordinary figure gives a much greater effect to his acting. He kept the whole theatre in roars of laughter. I liked Madame Blasis very much, but nothing will do after Malibran, who sung some comic songs, and [a] person's heart must have been made of stone not to have lost it to her. I lodged very near the Wedgwoods, and lived entirely with them, which was very pleasant, and had you been there it would have been quite perfect. It knocked me up most dreadfully, and I will never attempt again to do two things the same day.

*   *   *   *   *

C. Darwin to W. D. Fox.

[Cambridge] Thursday [March, 1830].

MY DEAR FOX,

I am through my Little-Go ! ! ! I am too much exalted to humble myself by apologising for not having written before. But I assure you before I went in, and when my nerves were in a shattered and weak condition, your injured person often rose before my eyes and taunted me with my idleness. But I am through, through, through. I could write the whole sheet full with this delightful word. I went in yesterday, and have

* At Birmingham.   † De Begnis's Christian name was Giuseppe.

.just heard the joyful news. I shall not know for a week
which class I am in. The whole examination is carried on in
a different system. It has one grand advantage—being over
in one day. They are rather strict, and ask a wonderful
number of questions.

And now I want to know something about your plans ;
of course you intend coming up here : what fun we will have
together ; what beetles we will catch ; it will do my heart
good to go once more together to some of our old haunts.   I
have two very promising pupils in Entomology, and we will
make regular campaigns into the Fens.   Heaven protect the
beetles and Mr. Jenyns, for we won't leave him a pair in the
whole country.   My new Cabinet is come down, and a gay
little affair it is.

And now for the time—I think I shall go for a few days
to town to hear an opera and see Mr. Hope ; not to mention
my brother also, whom I should have no objection to see.
If I go pretty soon, you can come afterwards, but if you will
settle your plans definitely, I will arrange mine, so send me a
letter by return of post.   And I charge you let it be favourable—that is to say, come directly.   Holden has been
ordained, and drove the Coach out on the Monday.   I do not
think he is looking very well.   Chapman wants you and
myself to pay him a visit when you come up, and begs to be
remembered to you.   You must excuse this short letter, as
I have no end more to send off by this day's post.   I long to
see you again, and till then,

My dear good old Fox,

Yours most sincerely,

C. DARWIN.

[In August he was in North Wales and wrote to Fox :—
" I have been intending to write every hour for the last
fortnight, but *really* have had no time. I left Shrewsbury
this day fortnight ago, and have since that time been

working from morning to night in catching fish or beetles. This is literally the first idle day I have had to myself; for on the rainy days I go fishing, on the good ones entomologising. You may recollect that for the fortnight previous to all this, you told me not to write, so that I hope I have made out some sort of defence for not having sooner answered your two long and very agreeable letters."]

C. Darwin to W. D. Fox.

[Cambridge, November 5, 1830.]

MY DEAR FOX,

I have so little time at present, and am so disgusted by reading that I have not the heart to write to anybody. I have only written once home since I came up. This must excuse me for not having answered your three letters, for which I am really very much obliged. . . .

I have not stuck an insect this term, and scarcely opened a case. If I had time I would have sent you the insects which I have so long promised; but really I have not spirits or time to do anything. Reading makes me quite desperate; the plague of getting up all my subjects is next thing to intolerable. Henslow is my tutor, and a most *admirable* one he makes; the hour with him is the pleasantest in the whole day. I think he is quite the most perfect man I ever met with. I have been to some very pleasant parties there this term. His good-nature is unbounded.

I am sure you will be sorry to hear poor old Whitley's father is dead. In a worldly point of view it is of great consequence to him, as it will prevent him going to the Bar for some time.—(Be sure answer this:) What did you pay for the iron hoop you had made in Shrewsbury? Because I do not mean to pay the whole of the Cambridge man's bill. You need not trouble yourself about the Phallus, as I have bought up both species. I have heard men say that Henslow has some curious religious opinions. I never perceived anything of it, have you? I am very glad to hear, after all your delays, you have heard of a curacy where you may read all the commandments without endangering your throat. I am also still more glad to hear that your mother continues steadily to improve. I do trust that you will have no further cause for uneasiness. With every wish for your happiness, my dear old Fox,

Believe me yours most sincerely,

CHARLES DARWIN.

C. Darwin to W. D. Fox.

Cambridge, Sunday, January 23, 1831.

MY DEAR FOX,

I do hope you will excuse my not writing before I took my degree. I felt a quite inexplicable aversion to write to anybody. But now I do most heartily congratulate you upon passing your examination, and hope you find your curacy comfortable. If it is my last shilling (I have not many), I will come and pay you a visit.

I do not know why the degree should make one so miserable, both before and afterwards. I recollect you were sufficiently wretched before, and I can assure [you] I am now, and what makes it the more ridiculous is, I know not what about. I believe it is a beautiful provision of nature to make one regret the less leaving so pleasant a place as Cambridge ; and amongst all its pleasures—I say it for once and for all— none so great as my friendship with you. I sent you a newspaper yesterday, in which you will see what a good place [10th] I have got in the Poll. As for Christ's, did you ever see such a college for producing Captains and Apostles ?* There are no men either at Emmanuel or Christ's plucked. Cameron is

---

* The " Captain " is at the head of the " Poll " : the " Apostles " are the last twelve in the Mathematical Tripos.

gulfed, together with other three Trinity scholars ! My plans are not at all settled. I think I shall keep this term, and then go and economise at Shrewsbury, return and take my degree.

A man may be excused for writing so much about himself when he has just passed the examination ; so you must excuse [me]. And on the same principle do you write a letter brimful of yourself and plans. I want to know something about your examination. Tell me about the state of your nerves ; what books you got up, and how perfect. I take an interest about that sort of thing, as the time will come when I must suffer. Your tutor, Thompson, begged to be remembered to you, and so does Whitley. If you will answer this, I will send as many stupid answers as you can desire.

Believe me, dear Fox,

CHAS. DARWIN.

CHAPTER V.

THE APPOINTMENT TO THE 'BEAGLE.'

[IN a letter addressed to Captain Fitz-Roy, before the *Beagle* sailed, my father wrote, " What a glorious day the 4th of November * will be to me—my second life will then commence, and it shall be as a birthday for the rest of my life."

The circumstances which led to this second birth—so much more important than my father then imagined—are connected with his Cambridge life, but may be more appropriately told in the present chapter. Foremost in the chain of circumstances which led to his appointment to the *Beagle*, was my father's friendship with Professor Henslow. He wrote in a pocket-book or diary, which contains a brief record of dates, &c., throughout his life :—

" 1831. *Christmas.*—Passed my examination for B.A. degree and kept the two following terms.

" During these months lived much with Professor Henslow, often dining with him and walking with him ; became slightly acquainted with several of the learned men in Cambridge, which much quickened the zeal which dinner parties and hunting had not destroyed.

" In the spring paid Mr. Dawes a visit with Ramsay and Kirby, and talked over an excursion to Teneriffe. In the spring Henslow persuaded me to think of Geology, and introduced me to Sedgwick. During Midsummer geologized a little in Shropshire.

* The *Beagle* did not however make her final and successful start until December 27.

"*August.* — Went on Geological tour* by Llangollen, Ruthin, Conway, Bangor, and Capel Curig, where I left Professor Sedgwick, and crossed the mountain to Barmouth."

In a letter to Fox (May 1831), my father writes :—" I am very busy . . . and see a great deal of Henslow, whom I do not know whether I love or respect most." His feeling for this admirable man is finely expressed in a letter which he wrote to Rev. L. Blomefield (then Rev. L. Jenyns), when the latter was engaged in his 'Memoir of Professor Henslow' (published 1862). The passage † has been made use of in the first of the memorial notices written for 'Nature,' and Mr. Romanes points out that my father, "while describing the character of another, is unconsciously giving a most accurate description of his own " :—

" I went to Cambridge early in the year 1828, and soon became acquainted, through some of my brother entomologists, with Professor Henslow, for all who cared for any branch of natural history were equally encouraged by him. Nothing could be more simple, cordial, and unpretending than the encouragement which he afforded to all young naturalists. I soon became intimate with him, for he had a remarkable power of making the young feel completely at ease with him ; though we were all awe-struck with the amount of his knowledge. Before I saw him, I heard one young man sum up his attainments by simply saying that he knew everything. When I reflect how immediately we felt at perfect ease with a man older, and in every way so immensely our superior, I think it was as much owing to the transparent sincerity of his character as to his kindness of heart ; and, perhaps, even still more, to a highly remarkable absence in him of all self-consciousness. One perceived at once that he never thought of

* Mentioned by Sedgwick in his preface to Salter's 'Catalogue of Cambrian and Silurian Fossils,' 1873.

† 'Memoir of the Rev. John Stevens Henslow, M.A.,' by the Rev. Leonard Jenyns. 8vo. London, 1862, p. 51.

his own varied knowledge or clear intellect, but solely on the
subject in hand. Another charm, which must have struck
every one, was that his manner to old and distinguished
persons and to the youngest student was exactly the same :
and to all he showed the same winning courtesy. He would
receive with interest the most trifling observation in any
branch of natural history ; and however absurd a blunder one
might make, he pointed it out so clearly and kindly, that one
left him no way disheartened, but only determined to be
more accurate the next time. In short, no man could be
better formed to win the entire confidence of the young, and
to encourage them in their pursuits.

"His Lectures on Botany were universally popular, and as
clear as daylight. So popular were they, that several of the
older members of the University attended successive courses.
Once every week he kept open house in the evening, and all
who cared for natural history attended these parties, which,
by thus favouring inter-communication, did the same good in
Cambridge, in a very pleasant manner, as the Scientific
Societies do in London. At these parties many of the most
distinguished members of the University occasionally attended ;
and when only a few were present, I have listened to the
great men of those days, conversing on all sorts of subjects,
with the most varied and brilliant powers. This was no small
advantage to some of the younger men, as it stimulated their
mental activity and ambition. Two or three times in each
session he took excursions with his botanical class ; either a
long walk to the habitat of some rare plant, or in a barge
down the river to the fens, or in coaches to some more distant
place, as to Gamlingay, to see the wild lily of the valley, and
to catch on the heath the rare natter-jack. These excursions
have left a delightful impression on my mind. He was, on
such occasions, in as good spirits as a boy, and laughed as
heartily as a boy at the misadventures of those who chased
the splendid swallow-tail butterflies across the broken and

treacherous fens. He used to pause every now and then and lecture on some plant or other object ; and something he could tell us on every insect, shell, or fossil collected, for he had attended to every branch of natural history. After our day's work we used to dine at some inn or house, and most jovial we then were. I believe all who joined these excursions will agree with me that they have left an enduring impression of delight on our minds.

"As time passed on at Cambridge I became very intimate with Professor Henslow, and his kindness was unbounded ; he continually asked me to his house, and allowed me to accompany him in his walks. He talked on all subjects, including his deep sense of religion, and was entirely open. I owe more than I can express to this excellent man. . . .

"During the years when I associated so much with Professor Henslow, I never once saw his temper even ruffled. He never took an ill-natured view of any one's character, though very far from blind to the foibles of others. It always struck me that his mind could not be even touched by any paltry feeling of vanity, envy, or jealousy. With all this equability of temper and remarkable benevolence, there was no insipidity of character. A man must have been blind not to have perceived that beneath this placid exterior there was a vigorous and determined will. When principle came into play, no power on earth could have turned him one hair's-breadth. . . .

" Reflecting over his character with gratitude and reverence, his moral attributes rise, as they should do in the highest character, in pre-eminence over his intellect."

In a letter to Rev. L. Blomefield (Jenyns), May 24, 1862, my father wrote with the same feelings that he had expressed in his letters thirty years before :—

" I thank you most sincerely for your kind present of your Memoir of Henslow. I have read about half, and it has interested me much. I did not think that I could have venerated him more than I did ; but your book has even exalted his character in my eyes. From turning over the pages of the latter half, I should think your account would be invaluable to any clergyman who wished to follow poor dear Henslow's noble example. What an admirable man he was."

The geological work mentioned in the quotation from my father's pocket-book was doubtless of importance as giving him some practical experience, and perhaps of more importance in helping to give him some confidence in himself. In July of the same year, 1831, he was " working like a tiger " at Geology, and trying to make a map of Shropshire, but not finding it " as easy as I expected."

In writing to Henslow about the same time, he gives some account of his work :—

" I should have written to you some time ago, only I was determined to wait for the clinometer, and I am very glad to say I think it will answer admirably. I put all the tables in my bedroom at every conceivable angle and direction. I will venture to say I have measured them as accurately as any geologist going could do . . . . I have been working at so many things that I have not got on much with geology. I suspect the first expedition I take, clinometer and hammer in hand, will send me back very little wiser and a good deal more puzzled than when I started. As yet I have only indulged in hypotheses, but they are such powerful ones that I suppose, if they were put into action but for one day, the world would come to an end."

He was evidently most keen to get to work with Sedgwick, for he wrote to Henslow : " I have not heard from Professor Sedgwick, so I am afraid he will not pay the Severn formations a visit. I hope and trust you did your best to urge him."

My father has given in his Recollections some account of this Tour.

There too we read of the projected excursion to the

Canaries, of which slight mention occurs in letters to Fox and Henslow.

In April 1831 he writes to Fox: "At present I talk, think, and dream of a scheme I have almost hatched of going to the Canary Islands. I have long had a wish of seeing tropical scenery and vegetation, and, according to Humboldt, Teneriffe is a very pretty specimen." And again in May : "As for my Canary scheme, it is rash of you to ask questions ; my other friends most sincerely wish me there, I plague them so with talking about tropical scenery, &c. Eyton will go next summer, and I am learning Spanish."

Later on in the summer the scheme took more definite form, and the date seems to have been fixed for June 1832. He got information in London about passage-money, and in July was working at Spanish and calling Fox "un grandísimo lebron," in proof of his knowledge of the language ; which, however, he found " intensely stupid." But even then he seems to have had some doubts about his companions' zeal, for he writes to Henslow (July 27, 1831): "I hope you continue to fan your Canary ardour. I read and re-read Humboldt ; do you do the same ? I am sure nothing will prevent us seeing the Great Dragon Tree."

Geological work and Teneriffe dreams carried him through the summer, till on returning from Barmouth for the sacred 1st of September, he received the offer of appointment as Naturalist to the *Beagle*.

The following extract from the pocket-book will be a help in reading the letters :—

"Returned to Shrewsbury at end of August. Refused offer of voyage.

" *September.*—Went to Maer, returned with Uncle Jos. to Shrewsbury, thence to Cambridge. London.

"11*th.*—Went with Captain Fitz-Roy in steamer to Plymouth to see the *Beagle*.

"*22nd.*—Returned to Shrewsbury, passing through Cambridge.

"*October 2nd.*—Took leave of my home. Stayed in London.

"*24th.*—Reached Plymouth.

"*October and November.*—These months very miserable.

"*December 10th.*—Sailed, but were obliged to put back.

"*21st.*—Put to sea again, and were driven back.

"*27th.*—Sailed from England on our Circumnavigation."]

George Peacock * to J. S. Henslow.

7 Suffolk Street, Pall Mall East [1831].

MY DEAR HENSLOW,

Captain Fitz-Roy is going out to survey the southern coast of Tierra del Fuego, and afterwards to visit many of the South Sea Islands, and to return by the Indian Archipelago. The vessel is fitted out expressly for scientific purposes, combined with the survey; it will furnish, therefore, a rare opportunity for a naturalist, and it would be a great misfortune that it should be lost.

An offer has been made to me to recommend a proper person to go out as a naturalist with this expedition; he will be treated with every consideration. The Captain is a young man of very pleasing manners (a nephew of the Duke of Grafton), of great zeal in his profession, and who is very highly spoken of; if Leonard Jenyns could go, what treasures he might bring home with him, as the ship would be placed at his disposal whenever his inquiries made it necessary or desirable. In the absence of so accomplished a naturalist, is there any person whom you could strongly recommend? he must be such a person as would do credit to our recommenda-

* Formerly Dean of Ely, and Lowndean Professor of Astronomy at Cambridge.

tion. Do think of this subject, it would be a serious loss to the cause of natural science if this fine opportunity was lost.

*   *   *   *   *

The ship sails about the end of September.
Write immediately, and tell me what can be done.

Believe me,
My dear Henslow,
Most truly yours,
GEORGE PEACOCK.

J. S. Henslow to C. Darwin.

Cambridge, August 24, 1831.

MY DEAR DARWIN,

Before I enter upon the immediate business of this letter, let us condole together upon the loss of our inestimable friend poor Ramsay, of whose death you have undoubtedly heard long before this.

I will not now dwell upon this painful subject, as I shall hope to see you shortly, fully expecting that you will eagerly catch at the offer which is likely to be made you of a trip to Tierra del Fuego, and home by the East Indies. I have been asked by Peacock, who will read and forward this to you from London, to recommend him a Naturalist as companion to Captain Fitz-Roy, employed by Government to survey the southern extremity of America. I have stated that I consider you to be the best qualified person I know of who is likely to undertake such a situation. I state this not in the supposition of your being a *finished* naturalist, but as amply qualified for collecting, observing, and noting, anything worthy to be noted in Natural History. Peacock has the appointment at his disposal, and if he cannot find a man willing to take the office, the opportunity will probably be lost. Captain Fitz-Roy wants a man (I understand) more as a companion than a mere collector, and would not take any one, however good a naturalist, who was not recommended to him likewise as a *gentleman*. Particulars of salary, &c., I know nothing. The voyage is to last two years, and if you take plenty of books with you, anything you please may be done. You will have ample opportunities at command. In short, I suppose there never was a finer chance for a man of zeal and spirit; Captain Fitz-Roy is a young man. What I wish you to do is instantly to come and consult with Peacock (at No. 7 Suffolk Street, Pall Mall East, or else at the University Club), and learn further particulars. Don't put on any modest doubts or fears about your disqualifications, for I assure you I think you are the very man they are in search of; so conceive yourself to be tapped on the shoulder by your bum-bailiff and affectionate friend,

J. S. HENSLOW.

The expedition is to sail on 25th September (at earliest), so there is no time to be lost.

G. Peacock to C. Darwin.

[1831.]

MY DEAR SIR,

I received Henslow's letter last night too late to forward it to you by the post; a circumstance which I do not regret, as it has given me an opportunity of seeing Captain Beaufort at the Admiralty (the Hydrographer), and of stating to him the offer which I have to make to you. He entirely approves of it, and you may consider the situation as at your absolute disposal. I trust that you will accept it, as it is an opportunity which should not be lost, and I look forward with great interest to the benefit which our collections of Natural History may receive from your labours.

The circumstances are these :—

Captain Fitz-Roy (a nephew of the Duke of Grafton) sails at the end of September, in a ship to survey, in the first instance, the South Coast of Tierra del Fuego, afterwards to visit the South Sea Islands, and to return by the Indian Archipelago to England. The expedition is entirely for scientific purposes, and the ship will generally wait your leisure for researches in Natural History, &c. Captain Fitz-Roy is a public-spirited and zealous officer, of delightful manners, and greatly beloved by all his brother officers. He went with Captain Beechey,* and spent £1500 in bringing over and educating at his own charge three natives of Patagonia. He engages at his own expense an artist at £200 a year to go with him. You may be sure, therefore, of having a very pleasant companion, who will enter heartily into all your views.

The ship sails about the end of September, and you must lose no time in making known your acceptance to Captain Beaufort, Admiralty Hydrographer. I have had a good deal of correspondence about this matter [with Henslow?], who feels, in common with myself, the greatest anxiety that you should go. I hope that no other arrangements are likely to interfere with it. . . .

The Admiralty are not disposed to give a salary, though they will furnish you with an official appointment, and every accommodation. If a salary should be required, however, I am inclined to think that it would be granted.

Believe me, my dear Sir,

Very truly yours,

GEORGE PEACOCK.

* For "Beechey," read "King." I do not find the name Fitz-Roy in the list of Beechey's officers. The Fuegians were brought back from Captain King's voyage.

C. Darwin to J. S. Henslow.

Shrewsbury, Tuesday [August 30, 1831].

MY DEAR SIR,

Mr. Peacock's letter arrived on Saturday, and I received it late yesterday evening. As far as my own mind is concerned, I should, I think *certainly*, most gladly have accepted the opportunity which you so kindly have offered me. But my father, although he does not decidedly refuse me, gives such strong advice against going, that I should not be comfortable if I did not follow it.

My father's objections are these: the unfitting me to settle down as a Clergyman, my little habit of seafaring, *the shortness of the time*, and the chance of my not suiting Captain Fitz-Roy. It is certainly a very serious objection, the very short time for all my preparations, as not only body but mind wants making up for such an undertaking. But if it had not been for my father I would have taken all risks. What was the reason that a Naturalist was not long ago fixed upon? I am very much obliged for the trouble you have had about it ; there certainly could not have been a better opportunity.

*　　*　　*　　*　　*

My trip with Sedgwick answered most perfectly. I did not hear of poor Mr. Ramsay's loss till a few days before your letter. I have been lucky hitherto in never losing any person for whom I had any esteem or affection. My acquaintance, although very short, was sufficient to give me those feelings in a great degree. I can hardly make myself believe he is no more. He was the finest character I ever knew.

Yours most sincerely,
My dear Sir,
CH. DARWIN.

I have written to Mr. Peacock, and I mentioned that I have asked you to send one line in the chance of his not getting my letter. I have also asked him to communicate with Captain Fitz-Roy. Even if I was to go, my father disliking would take away all energy, and I should want a good stock of that. Again I must thank you, it adds a little to the heavy but pleasant load of gratitude which I owe to you.

C. Darwin to R. W. Darwin.

[Maer] August 31 [1831].

MY DEAR FATHER,

I am afraid I am going to make you again very uncomfortable. But, upon consideration, I think you will excuse me once again, stating my opinions on the offer of the voyage. My excuse and reason is the different way all the Wedgwoods view the subject from what you and my sisters do.

I have given Uncle Jos* what I fervently trust is an accurate and full list of your objections, and he is kind enough to give his opinions on all. The list and his answers will be enclosed. But may I beg of you one favour, it will be doing me the greatest kindness, if you will send me a decided answer, yes or no? If the latter, I should be most ungrateful if I did not implicitly yield to your better judgment, and to the kindest indulgence you have shown me all through my life; and you may rely upon it I will never mention the subject again. If your answer should be yes; I will go directly to Henslow and consult deliberately with him, and then come to Shrewsbury.

The danger appears to me and all the Wedgwoods not great. The expense can not be serious, and the time I do not think, anyhow, would be more thrown away than if I stayed at home. But pray do not consider that I am so bent on going that I would for one *single moment* hesitate, if you thought that after a short period you should continue uncomfortable.

* Josiah Wedgwood.

I must again state I cannot think it would unfit me hereafter for a steady life. I do hope this letter will not give you much uneasiness. I send it by the car to-morrow morning; if you make up your mind directly will you send me an answer on the following day by the same means? If this letter should not find you at home, I hope you will answer as soon as you conveniently can.

I do not know what to say about Uncle Jos' kindness; I never can forget how he interests himself about me.

Believe me, my dear father,

Your affectionate son,

CHARLES DARWIN.

[Here follows the list of objections which are referred to in the following letter :—

(1.) Disreputable to my character as a Clergyman hereafter.

(2.) A wild scheme.

(3.) That they must have offered to many others before me the place of Naturalist.

(4.) And from its not being accepted there must be some serious objection to the vessel or expedition.

(5.) That I should never settle down to a steady life hereafter.

(6.) That my accommodations would be most uncomfortable.

(7.) That you [*i.e.* Dr. Darwin] should consider it as again changing my profession.

(8.) That it would be a useless undertaking.]

Josiah Wedgwood to R. W. Darwin.

Maer, August 31, 1831.
[Read this last.] *

MY DEAR DOCTOR,

I feel the responsibility of your application to me on the offer that has been made to Charles as being weighty, but as you have desired Charles to consult me, I cannot refuse to give the result of such consideration as I have been able to [give ?] it.

Charles has put down what he conceives to be your principal objections, and I think the best course I can take will be to state what occurs to me upon each of them.

1. I should not think that it would be in any degree disreputable to his character as a Clergyman. I should on the contrary think the offer honourable to him ; and the pursuit of Natural History, though certainly not professional, is very suitable to a clergyman.

2. I hardly know how to meet this objection, but he would have definite objects upon which to employ himself, and might acquire and strengthen habits of application, and I should think would be as likely to do so as in any way in which he is likely to pass the next two years at home.

3. The notion did not occur to me in reading the letters ; and on reading them again with that object in my mind I see no ground for it.

4. I cannot conceive that the Admiralty would send out a bad vessel on such a service. As to objections to the expedition, they will differ in each man's case, and nothing would, I think, be inferred in Charles's case, if it were known that others had objected.

5. You are a much better judge of Charles's character than I can be. If on comparing this mode of spending the next two years with the way in which he will probably spend

* In C. Darwin's writing.

them, if he does not accept this offer, you think him more likely to be rendered unsteady and unable to settle, it is undoubtedly a weighty objection. Is it not the case that sailors are prone to settle in domestic and quiet habits?

6. I can form no opinion on this further than that if appointed by the Admiralty he will have a claim to be as well accommodated as the vessel will allow.

7. If I saw Charles now absorbed in professional studies I should probably think it would not be advisable to interrupt them; but this is not, and, I think, will not be the case with him. His present pursuit of knowledge is in the same track as he would have to follow in the expedition.

8. The undertaking would be useless as regards his profession, but looking upon him as a man of enlarged curiosity, it affords him such an opportunity of seeing men and things as happens to few.

You will bear in mind that I have had very little time for consideration, and that you and Charles are the persons who must decide.

I am,

My dear Doctor,

Affectionately yours,

JOSIAH WEDGWOOD.

C. Darwin to J. S. Henslow.

Cambridge, Red Lion [Sept. 2], 1831.

MY DEAR SIR,

I am just arrived; you will guess the reason. My father has changed his mind. I trust the place is not given away.

I am very much fatigued, and am going to bed.

I dare say you have not yet got my second letter.

How soon shall I come to you in the morning? Send a verbal answer.

Good night,

Yours,

C. DARWIN.

C. Darwin to Miss Susan Darwin.

Cambridge, Sunday Morning [September 4, 1831].

MY DEAR SUSAN,

As a letter would not have gone yesterday, I put off writing till to-day. I had rather a wearisome journey, but got into Cambridge very fresh. The whole of yesterday I spent with Henslow, thinking of what is to be done, and that I find is a great deal. By great good luck I know a man of the name of Wood, nephew of Lord Londonderry. He is a great friend of Captain Fitz-Roy, and has written to him about me. I heard a part of Captain Fitz-Roy's letter, dated some time ago, in which he says : " I have a right good set of officers, and most of my men have been there before." It seems he has been there for the last few years ; he was then second in command with the same vessel that he has now chosen. He is only twenty-three years old, but [has] seen a deal of service, and won the gold medal at Portsmouth. The Admiralty say his maps are most perfect. He had choice of two vessels, and he chose the smallest. Henslow will give me letters to all travellers in town whom he thinks may assist me.

Peacock has sole appointment of Naturalist. The first person offered was Leonard Jenyns, who was so near accepting it that he packed up his clothes. But having two livings, he did not think it right to leave them—to the great regret of all his family. Henslow himself was not very far from accepting it, for Mrs. Henslow most generously, and without being asked, gave her consent ; but she looked so miserable that Henslow at once settled the point.

*   *   *   *   *

I am afraid there will be a good deal of expense at first. Henslow is much against taking many things ; it is [the] mistake all young travellers fall into. I write as if it was settled, but Henslow tells me *by no means* to make up my mind till I have had long conversations with Captains

Beaufort and Fitz-Roy. Good-bye. You will hear from me
constantly. Direct 17 Spring Gardens. *Tell nobody* in
Shropshire yet. Be sure not.

C. DARWIN.

I was so tired that evening I was in Shrewsbury that
I thanked none of you for your kindness half so much as
I felt.

Love to my father.

The reason I don't want people told in Shropshire : in
case I should not go, it will make it more flat.

C. Darwin to Miss S. Darwin.

17 Spring Gardens, Monday
[September 5, 1831].

I have so little time to spare that I have none to waste in
re-writing letters, so that you must excuse my bringing up
the other with me and altering it. The last letter was
written in the morning. In [the] middle of [the] day, Wood
received a letter from Captain Fitz-Roy, which I must say
was *most* straightforward and *gentlemanlike*, but so much
against my going, that I immediately gave up the scheme ;
and Henslow did the same, saying that he thought Peacock
has acted *very wrong* in misrepresenting things so much.

I scarcely thought of going to town, but here I am ; and
now for more details, and much more promising ones.
Captain Fitz-Roy is [in] town, and I have seen him ; it is no
use attempting to praise him as much as I feel inclined to do,
for you would not believe me. One thing I am certain,
nothing could be more open and kind than he was to me. It
seems he had promised to take a friend with him, who is in
office and cannot go, and he only received the letter five
minutes before I came in ; and this makes things much better
for me, as want of room was one of Fitz-Roy's greatest
objections. He offers me to go share in everything in his

cabin if I like to come, and every sort of accommodation that I can have, but they will not be numerous. He says nothing would be so miserable for him as having me with him if I was uncomfortable, as in a small vessel we must be thrown together, and thought it his duty to state everything in the worst point of view. I think I shall go on Sunday to Plymouth to see the vessel.

There is something most extremely attractive in his manners and way of coming straight to the point. If I live with him, he says I must live poorly—no wine, and the plainest dinners. The scheme is not certainly so good as Peacock describes. Captain Fitz-Roy advises me not [to] make up my mind quite yet, but that, seriously, he thinks it will have much more pleasure than pain for me. The vessel does not sail till the 10th of October. It contains sixty men, five or six officers, &c., but is a small vessel. It will probably be out nearly three years. I shall pay to mess the same as [the] Captain does himself, £30 per annum ; and Fitz-Roy says if I spend, including my outfitting, £500, it will be beyond the extreme. But now for still worse news. The round the world is not *certain*, but the chance most excellent. Till that point is decided, I will not be so. And you may believe, after the many changes I have made, that nothing but my reason shall decide me.

Fitz-Roy says the stormy sea is exaggerated ; that if I do not choose to remain with them, I can at any time get home to England, so many vessels sail that way, and that during bad weather (probably two months), if I like, I shall be left in some healthy, safe and nice country ; that I shall always have assistance ; that he has many books, all instruments, guns, at my service ; that the fewer and cheaper clothes I take the better. The manner of proceeding will just suit me. They anchor the ship, and then remain for a fortnight at a place. I have made Captain Beaufort perfectly understand me. He says if I start and do not go round the world,

I shall have good reason to think myself deceived. I am to
call the day after to-morrow, and, if possible, to receive more
certain instructions. The want of room is decidedly the
most serious objection ; but Captain Fitz-Roy (probably
owing to Wood's letter) seems determined to make me [as]
comfortable as he possibly can. I like his manner of proceeding. He asked me at once, "Shall you bear being told
that I want the cabin to myself?—when I want to be alone.
If we treat each other this way, I hope we shall suit ; if not,
probably we should wish each other at the devil."

We stop a week at [the] Madeira Islands, and shall see
most of [the] big cities in South America. Captain Beaufort is
drawing up the track through the South Sea. I am writing
in [a] great hurry ; I do not know whether you take interest
enough to excuse treble postage. I hope I am judging
reasonably, and not through prejudice, about Captain Fitz-
Roy ; if so, I am sure we shall suit. I dine with him to-day.
I could write [a] great deal more if I thought you liked it, and
I had at present time. There is indeed a tide in the affairs
of man, and I have experienced it, and I had *entirely* given it
up till one to-day.

Love to my father. Dearest Susan, good-bye.

CH. DARWIN.

C. Darwin to J. S. Henslow.

London, Monday [September 5, 1831].

MY DEAR SIR,

Gloria in excelsis is the most moderate beginning I
can think of. Things are more prosperous than I should
have thought possible. Captain Fitz-Roy is everything that is
delightful. If I was to praise half so much as I feel inclined,
you would say it was absurd, only once seeing him. I think
he really wishes to have me. He offers me to mess with
him, and he will take care I have such room as is possible
But about the cases he says I must limit myself ; but then he

thinks like a sailor about size. Captain Beaufort says I shall be upon the Boards, and then it will only cost me like other officers. Ship sails 10th of October. Spends a week at Madeira Islands; and then Rio de Janeiro. They all think most extremely probable, home by the Indian archipelago ; but till that is decided, I will not be so.

What has induced Captain Fitz-Roy to take a better view of the case is, that Mr. Chester, who was going as a friend,. cannot go, so that I shall have his place in every respect.

Captain Fitz-Roy has [a] good stock of books, many of which were in my list, and rifles, &c., so that the outfit will be much less expensive than I supposed.

The vessel will be out three years. I do not object so that my father does not. On Wednesday I have another interview with Captain Beaufort, and on Sunday most likely go with Captain Fitz-Roy to Plymouth. So I hope you will keep on thinking on the subject, and just keep memoranda of what may strike you. I will call most probably on Mr. Burchell and introduce myself. I am in lodgings at 17 Spring Gardens. You cannot imagine anything more pleasant, kind, and open than Captain Fitz-Roy's manners were to me. I am sure it will be my fault if we do not suit.

What changes I have had. Till one to-day I was building castles in the air about hunting foxes in Shropshire, now llamas in South America.

There is indeed a tide in the affairs of men. If you see Mr. Wood, remember me very kindly to him.

Good-bye.

My dear Henslow,

Your most sincere friend,

CHAS. DARWIN.

Excuse this letter in such a hurry.

C. Darwin to W. D. Fox.

17 Spring Gardens, London,
September 6, 1831.

*   *   *   *   *

Your letter gave me great pleasure. You cannot imagine how much your former letter annoyed and hurt me.* But, thank heaven, I firmly believe that it was my *own entire* fault in so interpreting your letter. I lost a friend the other day, and I doubt whether the moral death (as I then wickedly supposed) of our friendship did not grieve me as much as the real and sudden death of poor Ramsay. We have known each other too long to need, I trust, any more explanations. But I will mention just one thing—that on my death-bed, I think I could say I never uttered one insincere (which at the time I did not fully feel) expression about my regard for you. One thing more—the sending *immediately* the insects, on my honour, was an unfortunate coincidence. I forgot how you naturally would take them. When you look at them now, I hope no unkindly feelings will rise in your mind, and that you will believe that you have always had in me a sincere, and, I will add, an obliged friend. The very many pleasant minutes that we spent together in Cambridge rose like departed spirits in judgment against me. May we have many more such, will be one of my last wishes in leaving England. God bless you, dear old Fox. May you always be happy.

Yours truly,
CHAS. DARWIN.

I have left your letter behind, so do not know whether I direct right.

* He had misunderstood a letter of Fox's as implying a charge of falsehood.

C. Darwin to Miss Susan Darwin.

17 Spring Gardens, Tuesday.
[September 6, 1831.]

MY DEAR SUSAN,

Again I am going to trouble you. I suspect, if I keep on at this rate, you will sincerely wish me at Tierra del Fuego, or any other Terra, but England. First I will give my commissions. Tell Nancy to make me some twelve instead of eight shirts. Tell Edward to send me up in my carpet-bag (he can slip the key in the bag tied to some string), my slippers, a pair of lightish walking-shoes, my Spanish books, my new microscope (about six inches long and three or four deep), which must have cotton stuffed inside; my geological compass; my father knows that; a little book, if I have got it in my bedroom—'Taxidermy.' Ask my father if he thinks there would be any objection to my taking arsenic for a little time, as my hands are not quite well, and I have always observed that if I once get them well, and change my manner of living about the same time, they will generally remain well. What is the dose? Tell Edward my gun is dirty. What is Erasmus's direction? Tell me if you think there is time to write and to receive an answer before I start, as I should like particularly to know what he thinks about it. I suppose you do not know Sir J. Mackintosh's direction?

I write all this as if it was settled, but it is not more than it was, excepting that from Captain Fitz-Roy wishing me so much to go, and, from his kindness, I feel a predestination I shall start. I spent a very pleasant evening with him yesterday. He must be more than twenty-three years old; he is of a slight figure, and a dark but handsome edition of Mr. Kynaston, and, according to my notions, pre-eminently good manners. He is all for economy, excepting on one point—viz., fire-arms. He recommends me strongly to get a

case of pistols like his, which cost £60!! and never to go on
shore anywhere without loaded ones, and he is doubting about
a rifle; he says I cannot appreciate the luxury of fresh meat
here. Of course I shall buy nothing till everything is settled;
but I work all day long at my lists, putting in and striking
out articles. This is the first really cheerful day I have spent
since I received the letter, and it all is owing to the sort of
involuntary confidence I place in my *beau ideal* of a Captain.
We stop at Teneriffe. His object is to stop at as many
places as possible. He takes out twenty chronometers, and
it will be a "sin" not to settle the longitude. He tells me to
get it down in writing at the Admiralty that I have the free
choice to leave as soon and whenever I like. I dare say you
expect I shall turn back at the Madeira; if I have a morsel of
stomach left, I won't give up. Excuse my so often troubling
and writing: the one is of great utility, the other a great
amusement to me. Most likely I shall write to-morrow.
Answer by return of post. Love to my father, dearest
Susan.

C. DARWIN.

As my instruments want altering, send my things by the
'Oxonian' the same night.

C. Darwin to Miss Susan Darwin.

London, Friday Morning, September 9, 1831.

MY DEAR SUSAN,

I have just received the parcel. I suppose it was not
delivered yesterday owing to the Coronation. I am very much
obliged to my father, and everybody else. Everything is done
quite right. I suppose by this time you have received my
letter written next day, and I hope will send off the things.
My affairs remain *in statu quo*. Captain Beaufort says I am
on the books for victuals, and he thinks I shall have no
difficulty about my collections when I come home. But he is

too deep a fish for me to make him out. The only thing that now prevents me finally making up my mind, is the want of certainty about the South Sea Islands; although morally I have no doubt we should go there whether or no it is put in the instructions. Captain Fitz-Roy says I do good by plaguing Captain Beaufort, it stirs him up with a long pole. Captain Fitz-Roy says he is sure he has interest enough (particularly if this Administration is not everlasting—I shall soon turn Tory!), anyhow, even when out, to get the ship ordered home by whatever track he likes. From what Wood says, I presume the Dukes of Grafton and Richmond interest themselves about him. By the way, Wood has been of the greatest use to me; and I am sure his personal introduction of me inclined Captain Fitz-Roy to have me.

To explain things from the very beginning : Captain Fitz-Roy first wished to have a Naturalist, and then he seems to have taken a sudden horror of the chances of having somebody he should not like on board the vessel. He confesses his letter to Cambridge was to throw cold water on the scheme. I don't think we shall quarrel about politics, although Wood (as might be expected from a Londonderry) solemnly warned Fitz-Roy that I was a Whig. Captain Fitz-Roy was before Uncle Jos., he said, "now your friends will tell you a sea-captain is the greatest brute on the face of the creation. I do not know how to help you in this case, except by hoping you will give me a trial." How one does change ! I actually now wish the voyage was longer before we touch land. I feel my blood run cold at the quantity I have to do. Everybody seems ready to assist me. The Zoological want to make me a corresponding member. All this I can construe without crossing the Equator. But one friend is quite invaluable, viz. a Mr. Yarrell, a stationer, and excellent naturalist.* He goes

---

* William Yarrell, well known for his 'History of British Birds' and 'History of British Fishes,' was born in 1784. He inherited from his father a newsagent's business, to which he steadily adhered up to

to the shops with me and bullies about prices (not that I yet buy): hang me if I give £60 for pistols.

Yesterday all the shops were shut, so that I could do nothing ; and I was child enough to give £1. 1s. for an excellent seat to see the Procession.* And it certainly was very well worth seeing. I was surprised that any quantity of gold could make a long row of people quite glitter. It was like only what one sees in picture-books of Eastern processions. The King looked very well, and seemed popular, but there was very little enthusiasm ; so little that I can hardly think there will be a coronation this time fifty years.

The Life Guards pleased me as much as anything—they are quite magnificent ; and it is beautiful to see them clear a crowd. You think that they must kill a score at least, and apparently they really hurt nobody, but most deucedly frighten them. Whenever a crowd was so dense that the people were forced off the causeway, one of these six-feet gentlemen, on a black horse, rode straight at the place, making his horse rear very high, and fall on the thickest spot. You would suppose men were made of sponge to see them shrink away.

In the evening there was an illumination, and much grander than the one on the Reform Bill. All the principal streets were crowded just like a race-ground. Carriages generally being six abreast, and I will venture to say not going one mile an hour. The Duke of Northumberland learnt a lesson last time, for his house was very grand ; much more so than the other great nobility, and in much better taste ; every window in his house was full of straight lines of brilliant lights, and from their extreme regularity and number had a beautiful effect. The paucity of invention was very striking,

---

his death, "in his 73rd year." He was a man of a thoroughly amiable and honourable character, and was a valued office-bearer of several of the learned Societies.

* The Coronation of William IV.

crowns, anchors, and "W. R.'s" were repeated in endless succession. The prettiest were gas-pipes with small holes; they were almost painfully brilliant. I have written so much about the Coronation, that I think you will have no occasion to read the *Morning Herald.*

For about the first time in my life I find London very pleasant; hurry, bustle, and noise are all in unison with my feelings. And I have plenty to do in spare moments. I work at Astronomy, as I suppose it would astound a sailor if one did not know how to find Latitude and Longitude. I am now going to Captain Fitz-Roy, and will keep [this] letter open till evening for anything that may occur. I will give you one proof of Fitz-Roy being a good officer—all the officers are the same as before; two-thirds of his crew and [the] eight marines who went before all offered to come again, so the service cannot be so very bad. The Admiralty have just issued orders for a large stock of canister-meat and lemon-juice, &c. &c. I have just returned from spending a long day with Captain Fitz-Roy, driving about in his gig, and shopping. This letter is too late for to-day's post. You may consider it settled that I go. Yet there is room for change if any unto-ward accident should happen; this I can see no reason to expect. I feel convinced nothing else will alter my wish of going. I have begun to order things. I have procured a case of good strong pistols and an excellent rifle for £50, there is a saving; a good telescope, with compass, £5, and these are nearly the only expensive instruments I shall want. Captain Fitz-Roy has everything. I never saw so (what I should call, he says not) extravagant a man, as regards him-self, but as economical towards me. How he did order things! His fire-arms will cost £400 at least. I found the carpet bag when I arrived all right, and much obliged. I do not think I shall take any arsenic; shall send partridges to Mr. Yarrell; much obliged. Ask Edward to *bargain with* Clemson to make for my gun—*two spare* hammers or cocks, two main-springs, two sere-springs, four nipples or plugs—I mean one for each barrel, except nipples, of which there must be two for each, all of excellent quality, and set about them immediately; tell Edward to make inquiries about prices. I go on Sunday per packet to Plymouth, shall stay one or two days, then return, and hope to find a letter from you ; a few days in London ; then Cambridge, Shrewsbury, London, Plymouth, Madeira, is my route. It is a great bore my writing so much about the Coronation ; I could fill another sheet. I have just been with Captain King, Fitz-Roy's senior officer last expedition ; he thinks that the expedition will suit me. Unasked, he said Fitz-Roy's temper was perfect. He sends his own son with him as midshipman. The key of my microscope was forgotten ; it is of no consequence. Love to all.

CHAS. DARWIN.

C. Darwin to W. D. Fox.

17 Spring Gardens (and here I shall remain till I start)
[September 19, 1831].

MY DEAR FOX,

I returned from my expedition to see the *Beagle* at Plymouth on Saturday, and found your most welcome letter on my table. It is quite ridiculous what a very long period these last twenty days have appeared to me, certainly much more than as many weeks on ordinary occasions ; this will account for my not recollecting how much I told you of my plans.

*   *   *   *   *

But on the whole it is a grand and fortunate opportunity ; there will be so many things to interest me—fine scenery and an endless occupation and amusement in the different branches of Natural History ; then again navigation and meteorology will amuse me on the voyage, joined to the grand requisite of there being a pleasant set of officers, and, as far as I can judge, this is certain. On the other hand there is very considerable risk to one's life and health, and the leaving for so very long a time so many people whom I dearly love, is oftentimes a feeling so painful that it requires all my resolution to overcome it. But everything is now settled, and before the 20th of October I trust to be on the broad sea. My objection to the vessel is its smallness, which cramps one so for room for packing my own body and all my cases, &c. &c. As to its safety, I hope the Admiralty are the best judges; to a landsman's eye she looks very small. She is a ten-gun three-masted brig, but, I believe, an excellent vessel. So much for my future plans, and now for my present. I go to-night by the mail to Cambridge, and from thence, after settling my affairs, proceed to Shrewsbury (most likely on Friday 23rd, or perhaps before); there I shall stay a few days, and be in London by the 1st of October, and start for Plymouth on the 9th.

And now for the principal part of my letter. I do not know how to tell you how very kind I feel your offer of coming to see me before I leave England. Indeed I should like it very much; but I must tell you decidedly that I shall have very little time to spare, and that little time will be almost spoilt by my having so much to think about; and secondly, I can hardly think it worth your while to leave your parish for such a cause. But I shall never forget such generous kindness. Now I know you will act just as you think right; but do not come up for my sake. Any time is the same for me. I think from this letter you will know as much of my plans as I do myself, and will judge accordingly the where and when to write to me. Every now and then I have moments of glorious enthusiasm, when I think of the date and cocoa-trees, the palms and ferns so lofty and beautiful, everything new, everything sublime. And if I live to see years in after life, how grand must such recollections be! Do you know Humboldt? (if you don't, do so directly.) With what intense pleasure he appears always to look back on the days spent in the tropical countries. I hope when you next write to Osmaston, [you will] tell them my scheme, and give them my kindest regards and farewells.

Good-bye, my dear Fox,

Yours ever sincerely,

CHAS. DARWIN.

C. Darwin to R. Fitz-Roy.

17 Spring Gardens [October 17? 1831].

DEAR FITZ-ROY,

Very many thanks for your letter ; it has made me most comfortable, for it would have been heart-breaking to have left anything quite behind, and I never should have thought of sending things by some other vessel. This letter will, I trust, accompany some talc. I read your letter without attending to the name. But I have now procured some from Jones, which appears very good, and I will send it this evening by the mail. You will be surprised at not seeing me *proprià personà* instead of my handwriting. But I had just found out that the large steam-packet did not intend to sail on Sunday, and I was picturing to myself a small, dirty cabin, with the proportion of 39-40ths of the passengers very sick, when Mr. Earl came in and told me the *Beagle* would not sail till the beginning of November. This, of course, settled the point ; so that I remain in London one week more. I shall then send heavy goods by steamer and start myself by the coach on Sunday evening.

Have you a good set of mountain barometers? Several great guns in the scientific world have told me some points in geology to ascertain which entirely depend on their relative height. If you have not a good stock, I will add one more to the list. I ought to be ashamed to trouble you so much,

Q 2

but will you *send one line* to inform me? I am daily becoming more anxious to be off, and, if I am so, you must be
in a perfect fever. What a glorious day the 4th of November
will be to me! My second life will then commence, and it
shall be as a birthday for the rest of my life.

Believe me, dear Fitz-Roy,

Yours most sincerely,

CHAS. DARWIN.

Monday.—I hope I have not put you to much inconvenience by ordering the room in readiness.

C. Darwin to J. S. Henslow.

Devonport, November 15, 1831.

MY DEAR HENSLOW,

The orders are come down from the Admiralty, and
everything is finally settled. We positively sail the last day
of this month, and I think before that time the vessel will be
ready. She looks most beautiful, even a landsman must
admire her. *We* all think her the most perfect vessel ever
turned out of the Dockyard. One thing is certain, no vessel
has been fitted out so expensively, and with so much care.
Everything that can be made so is of mahogany, and nothing
can exceed the neatness and beauty of all the accommodations. The instructions are very general, and leave a great
deal to the Captain's discretion and judgment, paying a substantial as well as a verbal compliment to him.

*   *   *   *   *

No vessel ever left England with such a set of Chronometers, viz. twenty-four, all very good ones. In short, everything is well, and I have only now to pray for the sickness to
moderate its fierceness, and I shall do very well. Yet I
should not call it one of the very best opportunities for natural
history that has ever occurred. The absolute want of room is
an evil that nothing can surmount. I think L. Jenyns did

very wisely in not coming, that is judging from my own feelings, for I am sure if I had left college some few years, or been those years older, I *never* could have endured it. The officers (excepting the Captain) are like the freshest freshmen, that is in their manners, in everything else widely different. Remember me most kindly to him, and tell him if ever he dreams in the night of palm-trees, he may in the morning comfort himself with the assurance that the voyage would not have suited him.

I am much obliged for your advice, *de Mathematicis.* I suspect when I am struggling with a triangle, I shall often wish myself in your room, and as for those wicked sulky surds, I do not know what I shall do without you to conjure them. My time passes away very pleasantly. I know one or two pleasant people, foremost of whom is Mr. Thunder-and-lightning Harris,* whom I dare say you have heard of. My chief employment is to go on board the *Beagle*, and try to look as much like a sailor as I can. I have no evidence of having taken in man, woman or child.

I am going to ask you to do one more commission, and I trust it will be the last. When I was in Cambridge, I wrote to Mr. Ash, asking him to send my College account to my father, after having subtracted about £30 for my furniture. This he has forgotten to do, and my father has paid the bill, and I want to have the furniture-money transmitted to my father. Perhaps you would be kind enough to speak to Mr. Ash. I have cost my father so much money, I am quite ashamed of myself.

I will write once again before sailing, and perhaps you will write to me before then.

Remember me to Professor Sedgwick and Mr. Peacock.

Believe me, yours affectionately,

CHAS. DARWIN.

---

* William Snow Harris, the Electrician.

C. Darwin to J. S. Henslow.

Devonport, December 3, 1831.

MY DEAR HENSLOW,

It is now late in the evening, and to-night I am going to sleep on board. On Monday we most certainly sail, so you may guess in what a desperate state of confusion we are all in. If you were to hear the various exclamations of the officers, you would suppose we had scarcely had a week's notice. I am just in the same way taken all *aback*, and in such a bustle I hardly know what to do. The number of things to be done is infinite. I look forward even to sea-sickness with something like satisfaction, anything must be better than this state of anxiety. I am very much obliged for your last kind and affectionate letter. I always like advice from you, and no one whom I have the luck to know is more capable of giving it than yourself. Recollect, when you write, that I am a sort of *protégé* of yours, and that it is your bounden duty to lecture me.

I will now give you my direction: it is at first, Rio; but if you will send me a letter on the first Tuesday (when the packet sails) in February, directed to Monte Video, it will give me very great pleasure; I shall so much enjoy hearing a little Cambridge news. Poor dear old *Alma Mater!* I am a very worthy son in as far as affection goes. I have little more to write about . . . . I cannot end this without telling you how cordially I feel grateful for the kindness you have shown mè during my Cambridge life. Much of the pleasure and utility which I may have derived from it is owing to you. I long for the time when we shall again meet, and till then believe me, my dear Henslow,

Your affectionate and obliged friend,

CH. DARWIN.

Remember me most kindly to those who take any interest in me.

THE 'BEAGLE' LAID ASHORE, RIVER SANTA CRUZ.

CHAPTER VI.

THE VOYAGE.

"THERE is a natural good-humoured energy in his letters just like himself."—From a letter of Dr. R. W. Darwin's to Prof. Henslow.

[THE object of the *Beagle* voyage is briefly described in my father's 'Journal of Researches,' p. 1, as being "to complete the Survey of Patagonia and Tierra del Fuego, commenced under Captain King in 1826 to 1830; to survey the shores of Chile, Peru, and some islands in the Pacific; and to carry a chain of chronometrical measurements round the world."

The *Beagle* is described * as a well-built little vessel, of 235 tons, rigged as a barque, and carrying six guns. She belonged to the old class of ten-gun brigs, which were nick-named "coffins," from their liability to go down in severe weather. They were very "deep-waisted," that is, their bul-

* 'Voyages of the *Adventure* and *Beagle*,' vol. i. introduction xii. The illustration at the head of the chapter is from vol. ii. of the same work.

warks were high in proportion to their size, so that a heavy sea breaking over them might be highly dangerous. Nevertheless, she lived through the five years' work, in the most stormy regions in the world, under Commanders Stokes and Fitz-Roy without a serious accident. When re-commissioned in 1831 for her second voyage, she was found (as I learn from Admiral Sir James Sulivan) to be so rotten that she had practically to be rebuilt, and it was this that caused the long delay in refitting. The upper deck was raised, making her much safer in heavy weather, and giving her far more comfortable accommodation below. By these alterations and by the strong sheathing added to her bottom she was brought up to 242 tons burthen. It is a proof of the splendid seamanship of Captain Fitz-Roy and his officers that she returned without having carried away a spar, and that in only one of the heavy storms that she encountered was she in great danger.

She was fitted out for the expedition with all possible care, being supplied with carefully chosen spars and ropes, six boats, and a " dinghy ;" lightning conductors, "invented by Mr. Harris, were fixed in all the masts, the bowsprits, and even in the flying jib-boom." To quote my father's description, written from Devonport, November 17, 1831 : " Everybody, who can judge, says it is one of the grandest voyages that has almost ever been sent out. Everything is on a grand scale. . . . In short, everything is as prosperous as human means can make it." The twenty-four chronometers and the mahogany fittings seem to have been especially admired, and are again alluded to.

Owing to the smallness of the vessel, every one on board was cramped for room, and my father's accommodation seems to have been small enough : " I have just room to turn round," he writes to Henslow, " and that is all." Admiral Sir James Sulivan writes to me : " The narrow space at the end of the chart-table was his only accommodation for working, dressing, and sleeping ; the hammock being left hanging over his head by day, when the sea was at all rough, that he might lie on it with a book in his hand when he could not any longer sit at the table. His only stowage for clothes being several small drawers in the corner, reaching from deck to deck ; the top one being taken out when the hammock was hung up, without which there was not length for it, so then the foot-clews took the place of the top drawer. For specimens he had a very small cabin under the forecastle."

Yet of this narrow room he wrote enthusiastically, September 17, 1831 :—" When I wrote last I was in great alarm about my cabin. The cabins were not then marked out, but when I left they were, and mine is a capital one, certainly next best to the Captain's and remarkably light. My companion most luckily, I think, will turn out to be the officer whom I shall like best. Captain Fitz-Roy says he will take care that one corner is so fitted up that I shall be comfortable in it and shall consider it my home, but that also I shall have the run of his. My cabin is the drawing one ; and in the middle is a large table, on which we two sleep in hammocks. But for the first two months there will be no drawing to be done, so that it will be quite a luxurious room, and good deal larger than the Captain's cabin."

My father used to say that it was the absolute necessity of tidiness in the cramped space on the *Beagle* that helped 'to give him his methodical habits of working.' On the *Beagle*, too, he would say, that he learned what he considered the golden rule for saving time ; *i.e.*, taking care of the minutes.

Sir James Sulivan tells me that the chief fault in the outfit of the expedition was the want of a second smaller vessel to act as tender. This want was so much felt by Captain Fitz-Roy that he hired two decked boats to survey the coast of Patagonia, at a cost of £1100, a sum which he had to supply, although the boats saved several thousand pounds to the country. He afterwards bought a schooner to act as a tender, thus saving the country a further large amount. He was ultimately ordered to sell the schooner, and was compelled to bear the loss himself, and it was only after his death that some inadequate compensation was made for all the losses which he suffered through his zeal.

For want of a proper tender, much of the work had to be done in small open whale boats, which were sent away from the ship for weeks together, and this in a climate, where the crews were exposed to severe hardship from the almost constant rains, which sometimes continued for weeks together. The completeness of the equipment was also in other respects largely due to the public spirit of Captain Fitz-Roy. He provided at his own cost an artist, and a skilled instrument-maker to look after the chronometers.* Captain Fitz-Roy's wish was to take "some well-educated and scientific person" as his private guest, but this generous offer was only accepted by my father on condition of being allowed to pay a fair share of the expense of the Captain's table ; he was, moreover, on the ship's books for victuals.

In a letter to his sister (July 1832) he writes contentedly of his manner of life at sea :—" I do not think I have ever given you an account of how the day passes. We breakfast at eight o'clock. The invariable maxim is to throw away all politeness—that is, never to wait for each other, and bolt off the minute one has done eating, &c. At sea, when the weather is calm, I work at marine animals, with which the whole ocean abounds. If there is any sea up I am either sick or contrive to read some voyage or travels. At one we dine. You shore-going people are lamentably mistaken about the manner of living on board. We have never yet (nor shall we) dined off salt meat. Rice and peas and *calavanses* are excellent vegetables, and, with good bread, who could want more? Judge Alderson could not be more temperate, as nothing but water comes on the table. At five we have tea.

* Either one or both were on the books for victuals.

The midshipmen's berth have all their meals an hour before us, and the gun-room an hour afterwards."

The crew of the *Beagle* consisted of Captain Fitz-Roy, "Commander and Surveyor," two lieutenants, one of whom (the first lieutenant) was the late Captain Wickham, Governor of Queensland ; the present Admiral Sir James Sulivan, K.C.B., was the second lieutenant. Besides the master and two mates, there was an assistant-surveyor, the present Admiral Lort Stokes. There were also a surgeon, assistant-surgeon, two midshipmen, master's mate, a volunteer (1st class), purser, carpenter, clerk, boatswain, eight marines, thirty-four seamen, and six boys.

There are not, I believe, many survivors of my father's old ship-mates. Admiral Mellersh, Mr. Hamond, and Mr. Philip King, of the Legislative Council of Sydney, and Mr. Usborne, are among the number. Admiral Johnson died almost at the same time as my father.

He retained to the last a most pleasant recollection of the voyage of the *Beagle*, and of the friends he made on board her. To his children their names were familiar, from his many stories of the voyage, and we caught his feeling of friendship for many who were to us nothing more than names.

It is pleasant to know how affectionately his old companions remember him.

Sir James Sulivan remained, throughout my father's lifetime, one of his best and truest friends. He writes :—" I can confidently express my belief that during the five years in the *Beagle*, he was never known to be out of temper, or to say one unkind or hasty word *of* or *to* any one. You will therefore readily understand how this, combined with the admiration of his energy and ability, led to our giving him the name of 'the dear old Philosopher.'" * Admiral Mellersh

* His other nickname was " The Flycatcher." I have heard my father tell how he overheard the boatswain of the *Beagle* showing another boatswain over the ship, and pointing out the officers : " That's our first lieutenant ; that's our doctor ; that's our flycatcher."

writes to me:—"Your father is as vividly in my mind's eye as if it was only a week ago that I was in the *Beagle* with him ; his genial smile and conversation can never be forgotten by any who saw them and heard them. I was sent on two or three occasions away in a boat with him on some of his scientific excursions, and always looked forward to these trips with great pleasure, an anticipation that, unlike many others, was always realised. I think he was the only man I ever knew against whom I never heard a word said ; and as people when shut up in a ship for five years are apt to get cross with each other, that is saying a good deal. Certainly we were always so hard at work, we had no time to quarrel, but if we had done so, I feel sure your father would have tried (and have been successful) to throw oil on the troubled waters."

Admiral Stokes, Mr. King, Mr. Usborne, and Mr. Hamond, all speak of their friendship with him in the same warm-hearted way.

Of the life on board and on shore his letters give some idea. Captain Fitz-Roy was a strict officer, and made himself thoroughly respected both by officers and men. The occasional severity of his manner was borne with because every one on board knew that his first thought was his duty, and that he would sacrifice anything to the real welfare of the ship. My father writes, July 1834, "We all jog on very well together, there is no quarrelling on board, which is something to say. The Captain keeps all smooth by rowing every one in turn." The best proof that Fitz-Roy was valued as a commander is given by the fact that many * of the crew had sailed with him in the *Beagle's* former voyage, and there were a few officers as well as seamen and marines, who had served in the *Adventure* or *Beagle* during the whole of that expedition.

My father speaks of the officers as a fine determined set of

* 'Voyage of the *Adventure* and *Beagle*,' vol. ii. p. 21.

men, and especially of Wickham, the first lieutenant, as a
" glorious fellow." The latter being responsible for the smart-
ness and appearance of the ship strongly objected to his
littering the decks, and spoke of specimens as " d—d beastly
devilment," and used to add, " If I were skipper, I would soon
have you and all your d—d mess out of the place."

A sort of halo of sanctity was given to my father by the
fact of his dining in the Captain's cabin, so that the midshipmen used at first to call him " Sir," a formality, however,
which did not prevent his becoming fast friends with the
younger officers. He wrote about the year 1861 or 1862
to Mr. P. G. King, M.L.C., Sydney, who, as before stated,
was a midshipman on board the *Beagle* :—" The remembrance
of old days, when we used to sit and talk on the booms of
the *Beagle*, will always, to the day of my death, make me
glad to hear of your happiness and prosperity." Mr. King
describes the pleasure my father seemed to take " in pointing
out to me as a youngster the delights of the tropical nights,
with their balmy breezes eddying out of the sails above us,
and the sea lighted up by the passage of the ship through
the never-ending streams of phosphorescent animalculæ."

It has been assumed that his ill-health in later years was
due to his having suffered so much from sea-sickness. This
he did not himself believe, but rather ascribed his bad health
to the hereditary fault which came out as gout in some of
the past generations. I am not quite clear as to how much
he actually suffered from sea-sickness ; my impression is
distinct that, according to his own memory, he was not
actually ill after the first three weeks, but constantly uncomfortable when the vessel pitched at all heavily. But, judging
from his letters, and from the evidence of some of the officers,
it would seem that in later years he forgot the extent of the
discomfort from which he suffered. Writing June 3, 1836,
from the Cape of Good Hope, he says : " It is a lucky thing
for me that the voyage is drawing to its close, for I positively

suffer more from sea-sickness now than three years ago."
Admiral Lort Stokes wrote to the *Times*, April 25, 1883 :—

"May I beg a corner for my feeble testimony to the
marvellous persevering endurance in the cause of science of
that great naturalist, my old and lost friend, Mr. Charles
Darwin, whose remains are so very justly to be honoured
with a resting-place in Westminster Abbey?

"Perhaps no one can better testify to his early and most
trying labours than myself. We worked together for several
years at the same table in the poop cabin of the *Beagle*
during her celebrated voyage, he with his microscope and
myself at the charts. It was often a very lively end of
the little craft, and distressingly so to my old friend, who
suffered greatly from sea-sickness. After, perhaps, an hour's
work he would say to me, 'Old fellow, I must take the horizontal for it,' that being the best relief position from ship
motion; a stretch out on one side of the table for some time
would enable him to resume his labours for a while, when he
had again to lie down.

"It was distressing to witness this early sacrifice of Mr.
Darwin's health, who ever afterwards seriously felt the ill-
effects of the *Beagle's* voyage."

Mr. A. B. Usborne writes, " He was a dreadful sufferer
from sea-sickness, and at times, when I have been officer
of the watch, and reduced the sails, making the ship more
easy, and thus relieving him, I have been pronounced by him
to be 'a good officer,' and he would resume his microscopic
observations in the poop cabin." The amount of work that
he got through on the *Beagle* shows that he was habitually
in full vigour ; he, had, however, one severe illness in South
America, when he was received into the house of an English-
man, Mr. Corfield, who tended him with careful kindness.
I have heard him say that in this illness every secretion of
the body was affected, and that when he described the

symptoms to his father Dr. Darwin could make no guess as to the nature of the disease. My father was sometimes inclined to think that the breaking up of his health was to some extent due to this attack.

The *Beagle* letters give ample proof of his strong love of home, and all connected with it, from his father down to Nancy, his old nurse, to whom he sometimes sends his love.

His delight in home-letters is shown in such passages as :— " But if you knew the glowing, unspeakable delight, which I felt at being certain that my father and all of you were well, only four months ago, you would not grudge the labour lost in keeping up the regular series of letters."

Or again—his longing to return in words like these :— " It is too delightful to think that I shall see the leaves fall and hear the robin sing next autumn at Shrewsbury. My feelings are those of a schoolboy to the smallest point ; I doubt whether ever boy longed for his holidays as much as I do to see you all again. I am at present, although nearly half the world is between me and home, beginning to arrange what I shall do, where I shall go during the first week."

Another feature in his letters is the surprise and delight with which he hears of his collections and observations being of some use. It seems only to have gradually occurred to him that he would ever be more than a collector of specimens and facts, of which the great men were to make use. And even as to the value of his collections he seems to have had much doubt, for he wrote to Henslow in 1834: " I really began to think that my collections were so poor that you were puzzled what to say ; the case is now quite on the opposite tack, for you are guilty of exciting all my vain feelings to a most comfortable pitch ; if hard work will atone for these thoughts, I vow it shall not be spared."

After his return and settlement in London, he began to realise the value of what he had done, and wrote to Captain Fitz-Roy—" However others may look back to the *Beagle's* voyage, now that the small disagreeable parts are well-nigh forgotten, I think it far the *most fortunate circumstance in my life* that the chance afforded by your offer of taking a Naturalist fell on me.   I often have the most vivid and delightful pictures of what I saw on board the *Beagle* pass before my eyes.   These recollections, and what I learnt on Natural History, I would not exchange for twice ten thousand a year."

In selecting the following series of letters, I have been guided by the wish to give as much personal detail as possible.   I have given only a few scientific letters, to illustrate the way in which he worked, and how he regarded his own results.   In his ' Journal of Researches' he gives incidentally some idea of his personal character ; the letters given in the present chapter serve to amplify in fresher and more spontaneous words that impression of his personality which the ' Journal' has given to so many readers.]

C. Darwin to R. W. Darwin.

Bahia, or San Salvador, Brazils
[February 8, 1832].

MY DEAR FATHER,   I find after the first page I have been writing to my sisters.

I am writing this on the 8th of February, one day's sail past St. Jago (Cape de Verd), and intend taking the chance of meeting with a homeward-bound vessel somewhere about the equator.   The date, however, will tell this whenever the opportunity occurs.   I will now begin from the day of leaving England, and give a short account of our progress. We sailed, as you know, on the 27th of December, and have been fortunate enough to have had from that time to the present a fair and moderate breeze.   It afterwards proved that we had escaped a heavy gale in the Channel, another at Madeira, and another on [the] Coast of Africa.   But in escaping the gale, we felt its consequence—a heavy sea.   In

the Bay of Biscay there was a long and continuous swell, and
the misery I endured from sea-sickness is far beyond what I
ever guessed at. I believe you are curious about it. I will
give you all my dear-bought experience. Nobody who has
only been to sea for twenty-four hours has a right to say that
sea-sickness is even uncomfortable. The real misery only
begins when you are so exhausted that a little exertion makes
a feeling of faintness come on. I found nothing but lying in
my hammock did me any good. I must especially except
your receipt of raisins, which is the only food that the stomach
will bear.

On the 4th of January we were not many miles from
Madeira, but as there was a heavy sea running, and the
island lay to windward, it was not thought worth while to
beat up to it. It afterwards has turned out it was lucky we
saved ourselves the trouble. I was much too sick even to get
up to see the distant outline. On the 6th, in the evening, we
sailed into the harbour of Santa Cruz. I now first felt even
moderately well, and I was picturing to myself all the delights
of fresh fruit growing in beautiful valleys, and reading Humboldt's descriptions of the island's glorious views, when perhaps
you may nearly guess at our disappointment, when a small
pale man informed us we must perform a strict quarantine of
twelve days. There was a death-like stillness in the ship till
the Captain cried "up jib," and we left this long-wished for
place.

We were becalmed for a day between Teneriffe and the
Grand Canary, and here I first experienced any enjoyment.
The view was glorious. The Peak of Teneriffe was seen
amongst the clouds like another world. Our only drawback
was the extreme wish of visiting this glorious island. *Tell
Eyton never to forget either the Canary Islands or South
America;* that I am sure it will well repay the necessary
trouble, but that he must make up his mind to find a good
deal of the latter. I feel certain he will regret it if he does

not make the attempt. From Teneriffe to St. Jago the voyage was extremely pleasant. I had a net astern the vessel which caught great numbers of curious animals, and fully occupied my time in my cabin, and on deck the weather was so delightful and clear, that the sky and water together made a picture. On the 16th we arrived at Port Praya, the capital of the Cape de Verds, and there we remained twenty-three days, viz. till yesterday, the 7th of February. The time has flown away most delightfully, indeed nothing can be pleasanter; exceedingly busy, and that business both a duty and a great delight. I do not believe I have spent one half-hour idly since leaving Teneriffe. St. Jago has afforded me an exceedingly rich harvest in several branches of Natural History. I find the descriptions scarcely worth anything of many of the commoner animals that inhabit the Tropics. I allude, of course, to those of the lower classes.

Geologising in a volcanic country is most delightful; besides the interest attached to itself, it leads you into most beautiful and retired spots. Nobody but a person fond of Natural History can imagine the pleasure of strolling under cocoa-nuts in a thicket of bananas and coffee-plants, and an endless number of wild flowers. And this island, that has given me so much instruction and delight, is reckoned the most uninteresting place that we perhaps shall touch at during our voyage. It certainly is generally very barren, but the valleys are more exquisitely beautiful, from the very contrast. It is utterly useless to say anything about the scenery; it would be as profitable to explain to a blind man colours, as to a person who has not been out of Europe, the total dissimilarity of a tropical view. Whenever I enjoy anything, I always either look forward to writing it down, either in my log-book (which increases in bulk), or in a letter; so you must excuse raptures, and those raptures badly expressed. I find my collections are increasing wonderfully, and from Rio I think I shall be obliged to send a cargo home.

All the endless delays which we experienced at Plymouth have been most fortunate, as I verily believe no person ever went out better provided for collecting and observing in the different branches of Natural History. In a multitude of counsellors I certainly found good. I find to my great surprise that a ship is singularly comfortable for all sorts of work. Everything is so close at hand, and being cramped makes one so methodical, that in the end I have been a gainer. I already have got to look at going to sea as a regular quiet place, like going back to home after staying away from it. In short, I find a ship a very comfortable house, with everything you want, and if it was not for sea-sickness the whole world would be sailors. I do not think there is much danger of Erasmus setting the example, but in case there should be, he may rely upon it he does not know one-tenth of the sufferings of sea-sickness.

I like the officers much more than I did at first, especially Wickham, and young King and Stokes, and indeed all of them. The Captain continues steadily very kind, and does everything in his power to assist me. We see very little of each other when in harbour, our pursuits lead us in such different tracks. I never in my life met with a man who could endure nearly so great a share of fatigue. He works incessantly, and when apparently not employed, he is thinking. If he does not kill himself, he will during this voyage do a wonderful quantity of work. I find I am very well, and stand the little heat we have had as yet as well as anybody. We shall soon have it in real earnest. We are now sailing for Fernando Noronha, off the coast of Brazil, where we shall not stay very long, and then examine the shoals between there and Rio, touching perhaps at Bahia. I will finish this letter when an opportunity of sending it occurs.

February 26th.—About 280 miles from Bahia. On the 10th we spoke the packet *Lyra*, on her voyage to Rio. I sent a short letter by her, to be sent to England on [the] first opportunity. We have been singularly unlucky in not meeting with any homeward-bound vessels, but I suppose [at] Bahia we certainly shall be able to write to England. Since writing the first part of [this] letter nothing has occurred except crossing the Equator, and being shaved. This most disagreeable operation, consists in having your face rubbed with paint and tar, which forms a lather for a saw which represents the razor, and then being half drowned in a sail filled with salt water. About 50 miles north of the line we touched at the rocks of St. Paul; this little speck (about $\frac{1}{4}$ of a mile across) in the Atlantic has seldom been visited. It is totally barren, but is covered by hosts of birds; they were so unused to men that we found we could kill plenty with stones and sticks. After remaining some hours on the island, we returned on board with the boat loaded with our prey. From this we went to Fernando Noronha, a small island where the [Brazilians] send their exiles. The landing there was attended with so much difficulty owing [to] a heavy surf that the Captain determined to sail the next day after arriving. My one day on shore was exceedingly interesting, the whole island is one single wood so matted together by creepers that it is very difficult to move out of the beaten path. I find the Natural History of all these unfrequented spots most exceedingly interesting, especially the geology. I have written this much in order to save time at Bahia.

Decidedly the most striking thing in the Tropics is the novelty of the vegetable forms. Cocoa-nuts could well be imagined from drawings, if you add to them a graceful lightness which no European tree partakes of. Bananas and plantains are exactly the same as those in hothouses, the acacias or tamarinds are striking from the blueness of their foliage; but of the glorious orange trees, no description, no drawings, will give any just idea; instead of the sickly green of our oranges, the native ones exceed the Portugal laurel in the darkness of their tint, and infinitely exceed it in beauty of form. Cocoa-nuts, papaws, the light green bananas, and oranges, loaded with fruit, generally surround the more luxuriant villages. Whilst viewing such scenes, one feels the impossibility that any description should come near the mark, much less be overdrawn.

March 1st.—Bahia, or San Salvador. I arrived at this place on the 28th of February, and am now writing this letter after having in real earnest strolled in the forests of the new world. No person could imagine anything so beautiful as the ancient town of Bahia, it is fairly embosomed in a luxuriant wood of beautiful trees, and situated on a steep bank, and overlooks the calm waters of the great bay of All Saints. The houses are white and lofty, and, from the windows being narrow and long, have a very light and elegant appearance. Convents, porticos, and public buildings, vary the uniformity of the houses ; the bay is scattered over with large ships ; in short, and what can be said more, it is one of the finest views in the Brazils. But the exquisite glorious pleasure of walking amongst such flowers, and such trees, cannot be comprehended but by those who have experienced it. Although in so low a latitude the locality is not disagreeably hot, but at present it is very damp, for it is the rainy season. I find the climate as yet agrees admirably with me ; it makes me long to live quietly for some time in such a country. If you really want to have [an idea] of tropical countries, study Humboldt. Skip the scientific parts, and commence after leaving Teneriffe. My feelings amount to admiration the more I read him. Tell Eyton (I find I am writing to my sisters !) how exceedingly I enjoy America, and that I am sure it will be a great pity if he does not make a start.

This letter will go on the 5th, and I am afraid will be some time before it reaches you ; it must be a warning how in other parts of the world you may be a long time without hearing. A year might by accident thus pass. About the 12th we start for Rio, but we remain some time on the way in sounding the Albrolhos shoals. Tell Eyton as far as my experience goes let him study Spanish, French, drawing, and Humboldt. I do sincerely hope to hear of (if not to see him) in South America. I look forward to the letters in Rio— till each one is acknowledged, mention its date in the next.

We have beat all the ships in manœuvring, so much so that the commanding officer says, we need not follow his example ; because we do everything better than his great ship. I begin to take great interest in naval points, more especially now, as I find they all say we are the No. 1 in South America. I suppose the Captain is a most excellent officer. It was quite glorious to-day how we beat the *Samarang* in furling sails. It is quite a new thing for a " sounding ship " to beat a regular man-of-war ; and yet the *Beagle* is not at all a particular ship. Erasmus will clearly perceive it when he hears that in the night I have actually sat down in the sacred precincts of the quarter deck. You must excuse these queer letters, and recollect they are generally written in the evening after my day's work. I take more pains over my log-book, so that eventually you will have a good account of all the places I visit. Hitherto the voyage has answered *admirably* to me, and yet I am now more fully aware of your wisdom in throwing cold water on the whole scheme ; the chances are so numerous of turning out quite the reverse ; to such an extent do I feel this, that if my advice was asked by any person on a similar occasion, I should be very cautious in encouraging him. I have not time to write to anybody else, so send to Maer to let them know, that in the midst of the glorious tropical scenery, I do not forget how instrumental they were in placing me there. I will not rapturise again, but I give myself great credit in not being crazy out of pure delight.

Give my love to every soul at home, and to the Owens.

I think one's affections, like other good things, flourish and increase in these tropical regions.

The conviction that I am walking in the New World is

even yet marvellous in my own eyes, and I dare say it is little less so to you, the receiving a letter from a son of yours in such a quarter.

Believe me, my dear Father,

Your most affectionate son,

CHARLES DARWIN.

C. Darwin to W. D. Fox.

Botofogo Bay, near Rio de Janeiro,
May, 1832.

MY DEAR FOX,

I have delayed writing to you and all my other friends till I arrived here and had some little spare time. My mind has been, since leaving England, in a perfect *hurricane* of delight and astonishment, and to this hour scarcely a minute has passed in idleness. . . . .

At St. Jago my natural history and most delightful labours commenced. During the three weeks I collected a host of marine animals, and enjoyed many a good geological walk. Touching at some islands, we sailed to Bahia, and from thence to Rio, where I have already been some weeks. My collections go on admirably in almost every branch. As for insects, I trust I shall send a host of undescribed species to England. I believe they have no small ones in the collections, and here this morning I have taken minute Hydropori, Noterus, Colymbetes, Hydrophilus, Hydrobius, Gromius, &c. &c., as specimens of fresh-water beetles. I am entirely occupied with land animals, as the beach is only sand. Spiders and the adjoining tribes have perhaps given me, from their novelty, the most pleasure. I think I have already taken several new genera.

But Geology carries the day: it is like the pleasure of gambling. Speculating, on first arriving, what the rocks may be, I often mentally cry out 3 to 1 tertiary against primitive;

but the latter have hitherto won all the bets.   So much for
the grand end of my voyage : in other respects things are
equally flourishing.   My life, when at sea, is so quiet, that to
a person who can employ himself, nothing can be pleasanter ;
the beauty of the sky and brilliancy of the ocean together
make a picture.   But when on shore, and wandering in the
sublime forests, surrounded by views more gorgeous than even
Claude ever imagined, I enjoy a delight which none but those
who have experienced it can understand.   If it is to be done,
it must be by studying Humboldt.   At our ancient snug
breakfasts, at Cambridge, I little thought that the wide
Atlantic would ever separate us ; but it is a rare privilege that
with the body, the feelings and memory are not divided.   On
the contrary, the pleasantest scenes in my life, many of which
have been in Cambridge, rise from the contrast of the present,
the more vividly in my imagination.   Do you think any
diamond beetle will ever give me so much pleasure as our old
friend *crux major ?* . . . . It is one of my most constant
amusements to draw pictures of the past ; and in them I
often see you and poor little Fan.   Oh, Lord, and then old
Dash, poor thing !   Do you recollect how you all tormented
me about his beautiful tail ?

. . . . Think when you are picking insects off a hawthorn-
hedge on a fine May day (wretchedly cold, I have no doubt),
think of me collecting amongst pine-apples and orange-trees ;
whilst staining your fingers with dirty blackberries, think
and be envious of ripe oranges.   This is a proper piece of
bravado, for I would walk through many a mile of sleet, snow,
or rain to shake you by the hand.   My dear old Fox, God
bless you.   Believe me,

Yours very affectionately,

CHAS. DARWIN.

C. Darwin to J. S. Henslow.

Rio de Janeiro, May 18, 1832.

MY DEAR HENSLOW,

*   *   *   *   *

Till arriving at Teneriffe (we did not touch at Madeira) I was scarcely out of my hammock, and really suffered more than you can well imagine from such a cause. At Santa Cruz, whilst looking amongst the clouds for the Peak, and repeating to myself Humboldt's sublime descriptions, it was announced we must perform twelve days' strict quarantine. We had made a short passage, so " Up jib," and away for St. Jago. You will say all this sounds very bad, and so it was ; but from that to the present time it has been nearly one scene of continual enjoyment. A net over the stern kept me at full work till we arrived at St. Jago. Here we spent three most delightful weeks. The geology was pre-eminently interesting, and I believe quite new; there are some facts on a large scale of upraised coast (which is an excellent epoch for all the volcanic rocks to date from), that would interest Mr. Lyell.

One great source of perplexity to me is an utter ignorance whether I note the right facts, and whether they are of sufficient importance to interest others. In the one thing collecting I cannot go wrong. St. Jago is singularly barren, and produces few plants or insects, so that my hammer was my usual companion, and in its company most delightful hours I spent. On the coast I collected many marine animals, chiefly gasteropodous (I think some new). I examined pretty accurately a *Caryophyllea*, and, if my eyes are not bewitched, former descriptions have not the slightest resemblance to the animal. I took several specimens of an Octopus which possessed a most marvellous power of changing its colours, equalling any chameleon, and evidently accommodating the changes to the colour of the ground which it passed over.

Yellowish green, dark brown, and red, were the prevailing
colours ; this fact appears to be new, as far as I can find out.
Geology and the invertebrate animals will be my chief object
of pursuit through the whole voyage.

We then sailed for Bahia, and touched at the rock of
St. Paul. This is a serpentine formation. Is it not the only
island in the Atlantic which is not volcanic? We likewise
stayed a few hours at Fernando Noronha ; a tremendous surf
was running so that a boat was swamped, and the Captain
would not wait. I find my life on board when we are on blue
water most delightful, so very comfortable and quiet—it is
almost impossible to be idle, and that for me is saying a good
deal. Nobody could possibly be better fitted in every respect
for collecting than I am ; many cooks have not spoiled the
broth this time. Mr. Brown's little hints about microscopes,
&c., have been invaluable. I am well off in books, the ' Dic-
tionnaire Classique' *is most useful.* If you should think of
any thing or book that would be useful to me, if you would
write one line, E. Darwin, Wyndham Club, St. James's Street,
he will procure them, and send them with some other things
to Monte Video, which for the next year will be my head-
quarters.

Touching at the Abrolhos, we arrived here on April 4th,
when amongst others I received your most kind letter. You
may rely on it during the evening I thought of the many most
happy hours I have spent with you in Cambridge. I am now
living at Botofogo, a village about a league from the city, and
shall be able to remain a month longer. The *Beagle* has gone
back to Bahia, and will pick me up on its return. There is a
most important error in the longitude of South America, to
settle which this second trip has been undertaken. Our
chronometers, at least sixteen of them, are going superbly ;
none on record have ever gone at all like them.

A few days after arriving I started on an expedition of
150 miles to Rio Macao, which lasted eighteen days. Here I

first saw a tropical forest in all its sublime grandeur—nothing
but the reality can give any idea how wonderful, how magnificent the scene is.   If I was to specify any one thing I should
give the pre-eminence to the host of parasitical plants.   Your
engraving is exactly true, but underrates rather than exag-
gerates the luxuriance.   I never experienced such intense
delight.   I formerly admired Humboldt, I now almost adore
him ; he alone gives any notion of the feelings which are
raised in the mind on first entering the Tropics.   I am now
collecting fresh-water and land animals ; if what was told me
in London is true, viz. that there are no small insects in the
collections from the Tropics, I tell Entomologists to look out
and have their pens ready for describing.   I have taken as
minute (if not more so) as in England, Hydropori, Hygroti,
Hydrobii, Pselaphi, Staphylini, Curculio, &c. &c.   It is exceedingly interesting observing the difference of genera and
species from those which I know ; it is however much less
than I had expected.   I am at present red-hot with spiders ;
they are very interesting, and if I am not mistaken I have
already taken some new genera.   I shall have a large box to
send very soon to Cambridge, and with that I will mention
some more natural history particulars.

The Captain does everything in his power to assist me, and
we get on very well, but I thank my better fortune he has not
made me a renegade to Whig principles.   I would not be a
Tory, if it was merely on account of their cold hearts about
that scandal to Christian nations—Slavery.   I am very good
friends with all the officers.

I have just returned from a walk, and as a specimen, how
little the insects are known.   Noterus, according to the 'Dic-
tionnaire Classique,' contains solely three European species.
I in one haul of my net took five distinct species ; is this not
quite extraordinary? . . . .

Tell Professor Sedgwick he does not know how much
I am indebted to him for the Welsh Expedition ; it has

given me an interest in Geology which I would not give up for any consideration. I do not think I ever spent a more delightful three weeks than pounding the North-west Mountains. I look forward to the geology about Monte Video as I hear there are slates there, so I presume in that district I shall find the junctions of the Pampas, and the enormous granite formation of Brazils. At Bahia the pegmatite and gneiss in beds had the same direction, as observed by Humboldt, prevailing over Columbia, distant 1300 miles—is it not wonderful? Monte Video will be for a long time my direction. I hope you will write again to me, there is nobody from whom I like receiving advice so much as from you. . . . Excuse this almost unintelligible letter, and believe me, my dear Henslow, with the warmest feelings of respect and friendship,

Yours affectionately,
CHAS. DARWIN.

C. Darwin to J. M. Herbert.

Botofogo Bay, Rio de Janeiro,
June 1832.

MY DEAR OLD HERBERT,

Your letter arrived here when I had given up all hopes of receiving another, it gave me, therefore, an additional degree of pleasure. At such an interval of time and space one does learn to feel truly obliged to those who do not forget one. The memory when recalling scenes past by, affords to us *exiles* one of the greatest pleasures. Often and often whilst wandering amongst these hills do I think of Barmouth, and, I may add, as often wish for such a companion. What a contrast does a walk in these two places afford; here abrupt and stony peaks are to the very summit enclosed by luxuriant woods; the whole surface of the country, excepting where cleared by man, is one impenetrable forest. How different from Wales, with its sloping hills covered with turf, and its open valleys. I was not previously aware how intimately what may be called the moral part is connected with the enjoyment of scenery. I mean such ideas, as the history of the country, the utility of the produce, and more especially the happiness of the people living with them. Change the English labourer into a poor slave, working for another, and you will hardly recognise the same view. I am sure you will be glad to hear how very well every part (Heaven forefend, except sea-sickness) of the expedition has answered. We have already seen Teneriffe and the Great Canary; St. Jago, where I spent three most delightful weeks, revelling in the delights of first naturalising a tropical volcanic island, and besides other islands, the two celebrated ports in the Brazils, viz. Bahia and Rio.

I was in my hammock till we arrived at the Canaries, and I shall never forget the sublime impression the first view of Teneriffe made on my mind. The first arriving into warm weather was most luxuriously pleasant; the clear blue sky of the Tropics was no common change after those accursed south-west gales at Plymouth. About the Line it became weltering hot. We spent one day at St. Paul's, a little group of rocks about a quarter of a mile in circumference, peeping up in the midst of the Atlantic. There was such a scene here. Wickham (1st Lieutenant) and I were the only two who landed with guns and geological hammers, &c. The birds by myriads were too close to shoot; we then tried stones, but at last, *proh pudor!* my geological hammer was the instrument of death. We soon loaded the boat with birds and eggs. Whilst we were so engaged, the men in the boat were fairly fighting with the sharks for such magnificent fish as you could not see in the London market. Our boat would have made a fine subject for Snyders, such a medley of game it contained. We have been here ten weeks, and shall now start for Monte Video, when I look forward to many a gallop over the Pampas. I am ashamed of sending such a scrambling letter,

but if you were to see the heap of letters on my table, you
would understand the reason. . . .

I am glad to hear music flourishes so well in Cambridge;
but it [is] as barbarous to talk to me of "celestial concerts" as
to a person in Arabia of cold water.   In a voyage of this sort,
if one gains many new and great pleasures, on the other side
the loss is not inconsiderable.   How should you like to be
suddenly debarred from seeing every person and place, which
you have ever known and loved, for five years?   I do assure
you I am occasionally "taken aback" by this reflection; and
then for man or ship it is not so easy to right again.   Remember me most sincerely to the remnant of most excellent
fellows whom I have the good luck to know in Cambridge—
I mean Whitley and Watkins.   Tell Lowe I am even beneath
his contempt.   I can eat salt beef and musty biscuits for
dinner.   See what a fall man may come to!

My direction for the next year and a half will be Monte
Video.

God bless you, my very dear old Herbert.   May you
always be happy and prosperous is my most cordial wish.

Yours affectionately,

CHAS. DARWIN.

C. Darwin to F. Watkins.

Monte Video, River Plata,
August 18, 1832.

MY DEAR WATKINS,

I do not feel very sure you will think a letter from
one so far distant will be worth having; I write therefore on
the selfish principle of getting an answer.   In the different
countries we visit the entire newness and difference from
England only serves to make more keen the recollection of
its scenes and delights.   In consequence the pleasure of
thinking of, and hearing from one's former friends, does indeed
become great.   Recollect this, and some long winter's evening

sit down and send me a long account of yourself and
our friends ; both what you have, and what [you] intend
doing ; otherwise in three or four more years when I return
you will be all strangers to me. Considering how many
months have passed, we have not in the *Beagle* made much
way round the world. Hitherto everything has well repaid
the necessary trouble and loss of comfort. We stayed three
weeks at the Cape de Verds ; it was no ordinary pleasure
rambling over the plains of lava under a tropical sun, but
when I first entered on and beheld the luxuriant vegetation
in Brazil it was realizing the visions in the 'Arabian Nights.'
The brilliancy of the scenery throws one into a delirium of
delight, and a beetle hunter is not likely soon to awaken from
it, when whichever way he turns fresh treasures meet his eye.
At Rio de Janeiro three months passed away like so many
weeks. I made a most delightful excursion during this time
of 150 miles into the country. I stayed at an estate which
is the last of the cleared ground, behind is one vast impenetrable forest. It is almost impossible to imagine the quietude
of such a life. Not a human being within some miles in-
terrupts the solitude. To seat oneself amidst the gloom of
such a forest on a decaying trunk, and then think of home,
is a pleasure worth taking some trouble for.

We are at present in a much less interesting country.
One single walk over the undulatory turf plain shows everything which is to be seen. It is not at all unlike Cambridgeshire, only that every hedge, tree and hill must be levelled,
and arable land turned into pasture. All South America is
in such an unsettled state that we have not entered one port
without some sort of disturbance. At Buenos Ayres a shot
came whistling over our heads ; it is a noise I had never
before heard, but I found I had an instinctive knowledge of
what it meant. The other day we landed our men here, and
took possession at the request of the inhabitants of the central
fort. We philosophers do not bargain for this sort of work,

and I hope there will be no more. We sail in the course of a
day or two to survey the coast of Patagonia ; as it is entirely
unknown, I expect a good deal of interest. But already do
I perceive the grievous difference between sailing on these
seas and the Equinoctial ocean. In the " Ladies' Gulf," as the
Spaniards call it, it is so luxurious to sit on deck and enjoy
the coolness of the night, and admire the new constellations
of the South. . . . I wonder when we shall ever meet again ;
but be it when it may, few things will give me greater pleasure
than to see you again, and talk over the long time we have
passed together.

If you were to meet me at present I certainly should be
looked at like a wild beast, a great grizzly beard and flushing
jacket would disfigure an angel. Believe me, my dear Watkins, with the warmest feelings of friendship,

Ever yours,

CHARLES DARWIN.

C. Darwin to J. S. Henslow.

April 11, 1833.

MY DEAR HENSLOW,

We are now running up from the Falkland Islands to
the Rio Negro (or Colorado). The *Beagle* will proceed to
Monte Video ; but if it can be managed I intend staying at the
former place. It is now some months since we have been at
a civilised port ; nearly all this time has been spent in the
most southern part of Tierra del Fuego. It is a detestable
place ; gales succeed gales with such short intervals that it is
difficult to do anything. We were twenty-three days off
Cape Horn, and could by no means get to the westward.
The last and final gale before we gave up the attempt was
unusually severe. A sea stove one of the boats, and there
was so much water on the decks that every place was afloat ;
nearly all the paper for drying plants is spoiled, and half of
this curious collection.

We at last ran into harbour, and in the boats got to the west by the inland channels. As I was one of this party I was very glad of it. With two boats we went about 300 miles, and thus I had an excellent opportunity of geologising and seeing much of the savages. The Fuegians are in a more miserable state of barbarism than I had expected ever to have seen a human being. In this inclement country they are absolutely naked, and their temporary houses are like what children make in summer with boughs of trees. I do not think any spectacle can be more interesting than the first sight of man in his primitive wildness. It is an interest which cannot well be imagined until it is experienced. I shall never forget this when entering Good Success Bay—the yell with which a party received us. They were seated on a rocky point, surrounded by the dark forest of beech; as they threw their arms wildly round their heads, and their long hair streaming, they seemed the troubled spirits of another world. The climate in some respects is a curious mixture of severity and mildness; as far as regards the animal kingdom, the former character prevails; I have in consequence not added much to my collections.

The Geology of this part of Tierra del Fuego was, as indeed every place is, to me very interesting. The country is non-fossiliferous, and a common-place succession of granitic rocks and slates; attempting to make out the relation of cleavage, strata, &c., &c., was my chief amusement. The mineralogy, however, of some of the rocks will, I think, be curious from their resemblance to those of volcanic origin.

*   *   *   *   *

After leaving Tierra del Fuego we sailed to the Falklands. I forgot to mention the fate of the Fuegians whom we took back to their country. They had become entirely European in their habits and wishes, so much so that the younger one had forgotten his own language, and their countrymen paid but very little attention to them. We built houses for them and planted gardens, but by the time we return again on our passage round the Horn, I think it will be very doubtful how much of their property will be left unstolen.

. . . When I am sea-sick and miserable, it is one of my highest consolations to picture the future when we again shall be pacing together the roads round Cambridge. That day is a weary long way off. We have another cruise to make to Tierra del Fuego next summer, and then our voyage round the world will really commence. Captain Fitz-Roy has purchased a large schooner of 170 tons. In many respects it will be a great advantage having a consort—perhaps it may somewhat shorten our cruise, which I most cordially hope it may. I trust, however, that the Coral Reefs and various animals of the Pacific may keep up my resolution. Remember me most kindly to Mrs. Henslow and all other friends ; I am a true lover of Alma. Mater and all its inhabitants,

Believe me, my dear Henslow,

Your affectionate and most obliged friend,

CHARLES DARWIN.

C. Darwin to Miss C. Darwin.

Maldonado, Rio Plata, May 22, 1833.

. . . The following business piece is to my father. Having a servant of my own would be a really great addition to my comfort. For these two reasons : as at present the Captain has appointed one of the men always to be with me, but I do not think it just thus to take a seaman out of the ship ; and, secondly, when at sea I am rather badly off for any one to wait on me. The man is willing to be my servant, and all the expenses would be under £60 per annum. I have taught him to shoot and skin birds, so that in my main object he is very useful. I have now left England nearly a year and a half, and I find my expenses are not above

£200 per annum; so that, it being hopeless (from time) to write for permission, I have come to the conclusion that you would allow me this expense. But I have not yet resolved to ask the Captain, and the chances are even that he would not be willing to have an additional man in the ship. I have mentioned this because for a long time I have been thinking about it.

June.—I have just received a bundle more letters. I do not know how to thank you all sufficiently. One from Catherine, Feb. 8th, another from Susan, March 3rd, together with notes from Caroline and from my father; give my best love to my father. I almost cried for pleasure at receiving it; it was very kind thinking of writing to me. My letters are both few, short, and stupid in return for all yours; but I always ease my conscience by considering the Journal as a long letter. If I can manage it, I will, before doubling the Horn, send the rest. I am quite delighted to find the hide of the Megatherium has given you all some little interest in my employments. These fragments are not, however, by any means the most valuable of the geological relics. I trust and believe that the time spent in this voyage, if thrown away for all other respects, will produce its full worth in Natural History; and it appears to me the doing what *little* we can to increase the general stock of knowledge is as respectable an object of life as one can in any likelihood pursue. It is more the result of such reflections (as I have already said) than much immediate pleasure which now makes me continue the voyage, together with the glorious prospect of the future, when passing the Straits of Magellan, we have in truth the world before us. Think of the Andes, the luxuriant forest of Guayaquil, the islands of the South Sea, and New South Wales. How many magnificent and characteristic views, how many and curious tribes of men we shall see! What fine opportunities for geology and for studying the infinite host of living beings! Is not this a prospect to

keep up the most flagging spirit ? If I was to throw it away,
I don't think I should ever rest quiet in my grave. I
certainly should be a ghost and haunt the British Museum.

How famously the Ministers appear to be going on.
I always much enjoy political gossip and what you at home
think will, &c., &c., take place. I steadily read up the weekly
paper, but it is not sufficient to guide one's opinion ; and
I find it a very painful state not to be as obstinate as a pig
in politics. I have watched how steadily the general feeling,
as shown at elections, has been rising against Slavery. What
a proud thing for England if she is the first European nation
which utterly abolishes it! I was told before leaving
England that after living in slave countries all my opinions
would be altered ; the only alteration I am aware of is
forming a much higher estimate of the negro character.
It is impossible to see a negro and not feel kindly towards
him ; such cheerful, open, honest expressions and such fine
muscular bodies. I never saw any of the diminutive Portu-
guese, with their murderous countenances, without almost
wishing for Brazil to follow the example of Hayti ; and,
considering the enormous healthy-looking black population,
it will be wonderful if, at some future day, it does not take
place. There is at Rio a man (I know not his title) who has
a large salary to prevent (I believe) the landing of slaves ; he
lives at Botofogo, and yet that was the bay where, during my
residence, the greater number of smuggled slaves were landed.
Some of the Anti-Slavery people ought to question about his
office ; it was the subject of conversation at Rio amongst
the lower English. . . . .

C. Darwin to J. M. Herbert.

Maldonado, Rio Plata, June 2, 1833.

MY DEAR HERBERT,

I have been confined for the last three days to a
miserable dark room, in an old Spanish house, from the torrents

of rain : I am not, therefore, in very good trim for writing ; but, defying the blue devils, I will send you a few lines, if it is merely to thank you very sincerely for writing to me. I received your letter, dated December 1st, a short time since. We are now passing part of the winter in the Rio Plata, after having had a hard summer's work to the south. Tierra del Fuego is indeed a miserable place ; the ceaseless fury of the gales is quite tremendous. One evening we saw old Cape Horn, and three weeks afterwards we were only thirty miles to windward of it. It is a grand spectacle to see all nature thus raging ; but Heaven knows every one in the *Beagle* has seen enough in this one summer to last them their natural lives.

The first place we landed at was Good Success Bay. It was here Banks and Solander met such disasters on ascending one of the mountains. The weather was tolerably fine, and I enjoyed some walks in a wild country, like that behind Barmouth. The valleys are impenetrable from the entangled woods, but the higher parts, near the limits of perpetual snow, are bare. From some of these hills the scenery, from its savage, solitary character, was most sublime. The only inhabitant of these heights is the guanaco, and with its shrill neighing it often breaks the stillness. The consciousness that no European foot had ever trod much of this ground added to the delight of these rambles. How often and how vividly have many of the hours spent at Barmouth come before my mind! I look back to that time with no common pleasure ; at this moment I can see you seated on the hill behind the inn, almost as plainly as if you were really there. It is necessary to be separated from all which one has been accustomed to, to know how properly to treasure up such recollections, and at this distance, I may add, how properly to esteem such as yourself, my dear old Herbert. I wonder when I shall ever see you again. I hope it may be, as you say, surrounded with heaps of parchment ; but then there must be,

sooner or later, a dear little lady to take care of you and your
house.   Such a delightful vision makes me quite envious.
This is a curious life for a regular shore-going person such as
myself ; the worst part of it is its extreme length. There is
certainly a great deal of high enjoyment, and on the contrary
a tolerable share of vexation of spirit. Everything, however,
shall bend to the pleasure of grubbing up old bones, and cap-
tivating new animals. By the way, you rank my Natural
History labours far too high. I am nothing more than a lions'
provider : I do not feel at all sure that they will not growl and
finally destroy me.

It does one's heart good to hear how things are going on in
England.   Hurrah for the honest Whigs ! I trust they will
soon attack that monstrous stain on our boasted liberty, Colonial Slavery.   I have seen enough of slavery and the dispositions of the negroes, to be thoroughly disgusted with the
lies and nonsense one hears on the subject in England.
Thank God, the cold-hearted Tories, who, as J. Mackintosh
used to say, have no enthusiasm, except against enthusiasm,
have for the present run their race.   I am sorry, by your
letter, to hear you have not been well, and that you partly
attribute it to want of exercise. I wish you were here amongst
the green plains ; we would take walks which would rival the
Dolgelly ones, and you should tell stories, which I would
believe, even to a *cubic fathom of pudding.*   Instead, I must
take my solitary ramble, think of Cambridge days, and pick up
snakes, beetles and toads. Excuse this short letter (you
know I never studied ' The Complete Letter-writer '), and
believe me, my dear Herbert,

Your affectionate friend,
CHARLES DARWIN.

C. Darwin to J. S. Henslow.

East Falkland Island, March, 1834.

. . . . . I am quite charmed with Geology, but, like the wise animal between two bundles of hay, I do not know which to like the best ; the old crystalline group of rocks, or the softer and fossiliferous beds. When puzzling about stratification, &c., I feel inclined to cry " a fig for your big oysters, and your bigger megatheriums." But then when digging out some fine bones, I wonder how any man can tire his arms with hammering granite. By the way I have not one clear idea about cleavage, stratification, lines of upheaval. I have no books which tell me much, and what they do I cannot apply to what I see. In consequence I draw my own conclusions, and most gloriously ridiculous ones they are, I sometimes fancy. . . . Can you throw any light into my mind by telling me what relation cleavage and planes of deposition bear to each other ?

And now for my second *section*, Zoology. I have chiefly been employed in preparing myself for the South Sea by examining the polypi of the smaller Corallines in these latitudes. Many in themselves are very curious, and I think are quite undescribed ; there was one appalling one, allied to a *Flustra*, which I dare say I mentioned having found to the northward, where the cells have a movable organ (like a vulture's head, with a dilatable beak), fixed on the edge. But what is of more general interest is the unquestionable (as it appears to me) existence of another species of ostrich, besides the *Struthio rhea*. All the Gauchos and Indians state it is the case, and I place the greatest faith in their observations. I have the head, neck, piece of skin, feathers, and legs of one. The differences are chiefly in the colour of the feathers and scales on legs, being feathered below the knees, nidification, and geographical distribution. So much for what I have

lately done ; the prospect before me is full of sunshine, fine
weather, glorious scenery, the geology of the Andes, plains
abounding with organic remains (which perhaps I may have
the good luck to catch in the very act of moving), and lastly,
an ocean, its shores abounding with life, so that, if nothing
unforeseen happens, I will stick to the voyage, although for
what I can see this may last till we return a fine set of white-
headed old gentlemen. I have to thank you most cordially
for sending me the books. I am now reading the Oxford
'Report;' * the whole account of your proceedings is most
glorious ; you remaining in England cannot well imagine
how excessively interesting I find the reports. I am sure
from my own thrilling sensations when reading them, that
they cannot fail to have an excellent effect upon all those
residing in distant colonies, and who have little opportunity
of seeing the periodicals. My hammer has flown with re-
doubled force on the devoted blocks ; as I thought over
the eloquence of the Cambridge President, I hit harder
and harder blows. I hope to give my arms strength for the
Cordilleras. You will send me through Capt. Beaufort a copy
of the Cambridge 'Report.'

I have forgotten to mention that for some time past, and
for the future, I will put a pencil cross on the pill-boxes containing insects, as these alone will require being kept particularly dry ; it may perhaps save you some trouble. When
this letter will go I do not know, as this little seat of discord
has lately been embroiled by a dreadful scene of murder, and
at present there are more prisoners than inhabitants. If a
merchant vessel is chartered to take them to Rio, I will send
some specimens (especially my few plants and seeds). Remember me to all my Cambridge friends. I love and treasure
up every recollection of dear old Cambridge. I am much

---

* The second meeting of the   Oxford in 1832, the following year
British Association was held at   it was at Cambridge.

obliged to you for putting my name down to poor Ramsay's monument ; I never think of him without the warmest admiration. Farewell, my dear Henslow.

Believe me your most obliged and affectionate friend,
CHARLES DARWIN.

C. Darwin to Miss C. Darwin.

East Falkland Island, April 6, 1834.
MY DEAR CATHERINE,

When this letter will reach you I know not, but probably some man-of-war will call here before, in the common course of events, I should have another opportunity of writing.

* * * * *

After visiting some of the southern islands, we beat up through the magnificent scenery of the Beagle Channel to Jemmy Button's * country. We could hardly recognise poor Jemmy. Instead of the clean, well-dressed stout lad we left him, we found him a naked, thin, squalid savage. York and Fuegia had moved to their own country some months ago, the former having stolen all Jemmy's clothes. Now he had nothing except a bit of blanket round his waist. Poor Jemmy was very glad to see us, and, with his usual good feeling, brought several presents (otter-skins, which are most valuable to themselves) for his old friends. The Captain offered to take him to England, but this, to our surprise, he at once refused. In the evening his young wife came alongside and showed us the reason. He was quite contented. Last year, in the height of his indignation, he said "his country people no *sabe* nothing—damned fools"—now they were very good people, with *too* much to eat, and all the

---

* Jemmy Button, York Minster, and Fuegia Basket, were natives of Tierra del Fuego, brought to England by Captain Fitz-Roy in his former voyage, and restored to their country by him in 1832.

luxuries of life. Jemmy and his wife paddled away in their
canoe loaded with presents, and very happy. The most
curious thing is, that Jemmy, instead of recovering his
own language, has taught all his friends a little English.
" J. Button's canoe " and " Jemmy's wife come," " Give me
knife," &c., was said by several of them.

We then bore away for this island—this little miserable
seat of discord. We found that the Gauchos, under pretence
of a revolution, had murdered and plundered all the Englishmen whom they could catch, and some of their own countrymen. All the economy at home makes the foreign movements
of England most contemptible. How different from old Spain.
Here we, dog-in-the-manger fashion, seize an island, and leave
to protect it a Union Jack ; the possessor has, of course, been
murdered ; we now send a lieutenant with four sailors, without
authority or instructions. A man-of-war, however, ventured
to leave a party of marines, and by their assistance, and the
treachery of some of the party, the murderers have all been
taken, there being now as many prisoners as inhabitants.
This island must some day become a very important halting-
place in the most turbulent sea in the world. It is mid-way
between Australia and the South Sea to England ; between
Chili, Peru, &c., and the Rio Plata and the Rio de Janeiro.
There are fine harbours, plenty of fresh water, and good
beef. It would doubtless produce the coarser vegetables.
In other respects it is a wretched place. A little time
since, I rode across the island, and returned in four days.
My excursion would have been longer, but during the
whole time it blew a gale of wind, with hail and snow.
There is no fire-wood bigger than heath, and the whole
country is, more or less, an elastic peat-bog. Sleeping out
at night was too miserable work to endure it for all the
rocks in South America.

We shall leave this scene of iniquity in two or three days,
and go to the Rio de la Sta. Cruz. One of the objects is to

look at the ship's bottom. We struck rather heavily on an
unknown rock off Port Desire, and some of her copper is torn
off. After this is repaired the Captain has a glorious scheme ;
it is to go to the very head of this river, that is probably to the
Andes. It is quite unknown ; the Indians tell us it is two
or three hundred yards broad, and horses can nowhere ford it.
I cannot imagine anything more interesting. Our plans then
are to go to Port Famine, and there we meet the *Adventure*,
who is employed in making the Chart of the Falklands. This
will be in the middle of winter, so I shall see Tierra del Fuego
in her white drapery. We leave the straits to enter the Pacific
by the Barbara Channel, one very little known, and which passes
close to the foot of Mount Sarmiento (the highest mountain in
the south, excepting Mt. ! ! Darwin ! !). We then shall scud
away for Concepcion in Chili. I believe the ship must once
again steer southward, but if any one catches me there again,
I will give him leave to hang me up as a scarecrow for all
future naturalists. I long to be at work in the Cordilleras,
the geology of this side, which I understand pretty well is so
intimately connected with periods of violence in that great
chain of mountains. The future is, indeed, to me a brilliant
prospect. You say its very brilliancy frightens you ; but
really I am very careful ; I may mention as a proof, in all my
rambles I have never had any one accident or scrape. . . .
Continue in your good custom of writing plenty of gossip ; I
much like hearing all about all things. Remember me most
kindly to Uncle Jos, and to all the Wedgwoods. Tell Charlotte
(their married names sound downright unnatural) I should
like to have written to her, to have told her how well everything is going on ; but it would only have been a transcript of
this letter, and I have a host of animals at this minute surrounding me which all require embalming and numbering. I
have not forgotten the comfort I received that day at Maer,
when my mind was like a swinging pendulum. Give my best
love to my father. I hope he will forgive all my extrava-

gance, but not as a Christian, for then I suppose he would
send me no more money.

Good-bye, dear, to you, and all your goodly sisterhood.

Your affectionate brother,

CHAS. DARWIN.

My love to Nancy; * tell her, if she was now to see me
with my great beard, she would think I was some worthy
Solomon, come to sell the trinkets.

C. Darwin to C. Whitley.

Valparaiso, July 23, 1834.

MY DEAR WHITLEY,

I have long intended writing, just to put you in mind
that there is a certain hunter of beetles, and pounder of rocks,
still in existence. Why I have not done so before I know
not, but it will serve me right if you have quite forgotten me.
It is a very long time since I have heard any Cambridge news;
I neither know where you are living or what you are doing.
I saw your name down as one of the indefatigable guardians
of the eighteen hundred philosophers. I was delighted to
see this, for when we last left Cambridge you were at sad
variance with poor science; you seemed to think her a public
prostitute working for popularity. If your opinions are the
same as formerly, you would agree most admirably with
Captain Fitz-Roy,—the object of his most devout abhorrence
is one of the d—d scientific Whigs. As captains of men-of-
war are the greatest men going, far greater than kings or
schoolmasters, I am obliged to tell him everything in my
own favour. I have often said I once had a very good
friend, an out-and-out Tory, and we managed to get on very
well together. But he is very much inclined to doubt if
ever I really was so much honoured; at present we hear
scarcely anything about politics; this saves a great deal

* His old nurse.

of trouble, for we all stick to our former opinions rather more obstinately than before, and can give rather fewer reasons for doing so.

I do hope you will write to me: ('H.M.S. *Beagle,* S. American Station' will find me). I should much like to hear in what state you are both in body and mind. ¿ *Quién sabe?* as the people say here (and God knows they well may, for they do know little enough), if you are not a married man, and may be nursing, as Miss Austen says, little olive branches, little pledges of mutual affection. Eheu! Eheu! this puts me in mind of former visions of glimpses into futurity, where I fancied I saw retirement, green cottages, and white petticoats. What will become of me hereafter I know not; I feel like a ruined man, who does not see or care how to extricate himself. That this voyage must come to a conclusion my reason tells me, but otherwise I see no end to it. It is impossible not bitterly to regret the friends and other sources of pleasure one leaves behind in England; in place of it there is much solid enjoyment, some present, but more in anticipation, when the ideas gained during the voyage can be compared to fresh ones. I find in Geology a never-failing interest, as it has been remarked, it creates the same grand ideas respecting this world which Astronomy does for the universe. We have seen much fine scenery; that of the Tropics in its glory and luxuriance exceeds even the language of Humboldt to describe. A Persian writer could alone do justice to it, and if he succeeded he would in England be called the 'Grandfather of all liars.'

But I have seen nothing which more completely astonished me than the first sight of a savage. It was a naked Fuegian, his long hair blowing about, his face besmeared with paint. There is in their countenances an expression which I believe, to those who have not seen it, must be inconceivably wild. Standing on a rock he uttered tones and made gesticulations, than which the cries of domestic animals are far more intelligible.

When I return to England, you must take me in hand with respect to the fine arts. I yet recollect there was a man called Raffaelle Sanctus. How delightful it will be once again to see, in the Fitzwilliam, Titian's Venus. How much more then delightful to go to some good concert or fine opera. These recollections will not do. I shall not be able to-morrow to pick out the entrails of some small animal with half my usual gusto. Pray tell me some news about Cameron, Watkins, Marinden, the two Thompsons of Trinity, Lowe, Heaviside, Matthew. Herbert I have heard from. How is Henslow getting on? and all other good friends of dear Cambridge? Often and often do I think over those past hours, so many of which have been passed in your company. Such can never return, but their recollection can never die away.

God bless you, my dear Whitley,

Believe me, your most sincere friend,

CHAS. DARWIN.

C. Darwin to Miss C. Darwin.

Valparaiso, November 8, 1834.

MY DEAR CATHERINE,

My last letter was rather a gloomy one, for I was not very well when I wrote it. Now everything is as bright as sunshine. I am quite well again after being a second time in bed for a fortnight. Captain Fitz-Roy very generously has delayed the ship ten days on my account, and without at the time telling me for what reason.

We have had some strange proceedings on board the *Beagle*, but which have ended most capitally for all hands. Captain Fitz-Roy has for the last two months been working *extremely* hard, and at the same time constantly annoyed by interruptions from officers of other ships; the selling the schooner and its consequences were very vexatious; the cold manner the Admiralty (solely I believe because he is a Tory) have treated him, and a thousand other, &c. &c.'s, has made him very thin and unwell. This was accompanied by a morbid depression of spirits, and a loss of all decision and resolution. . . . All that Bynoe (the surgeon) could say, that it was merely the effect of bodily health and exhaustion after such application, would not do; he invalided, and Wickham was appointed to the command. By the instructions Wickham could only finish the survey of the southern part, and would then have been obliged to return direct to England. The grief on board the *Beagle* about the Captain's decision was universal and deeply felt; one great source of his annoyment was the feeling it impossible to fulfil the whole instructions; from his state of mind it never occurred to him that the very instructions order him to do as much of the West coast *as he has time for*, and then proceed across the Pacific.

Wickham (very disinterestedly giving up his own promotion) urged this most strongly, stating that when he took the command nothing should induce him to go to Tierra del Fuego again; and then asked the Captain what would be gained by his resignation? why not do the more useful part, and return as commanded by the Pacific. The Captain at last, to every one's joy, consented, and the resignation was withdrawn.

Hurrah! hurrah! it is fixed the *Beagle* shall not go one mile south of Cape Tres Montes (about 200 miles south of Chiloe), and from that point to Valparaiso will be finished in about five months. We shall examine the Chonos Archipelago, entirely unknown, and the curious inland sea behind Chiloe. For me it is glorious. Cape Tres Montes is the most southern point where there is much geological interest, as there the modern beds end. The Captain then talks of crossing the Pacific; but I think we shall persuade him to finish the Coast of Peru, where the climate is delightful, the country hideously sterile, but abounding with the highest interest to a geologist. For the first time since leaving England I now see a clear and not so distant prospect of returning to you all : crossing the Pacific, and from Sydney home, will not take much time. As soon as the Captain invalided I at once determined to leave the *Beagle*, but it was quite absurd what a revolution in five minutes was effected in all my feelings. I have long been grieved and most sorry at the interminable length of the voyage (although I never would have quitted it); but the minute it was all over, I could not make up my mind to return. I could not give up all the geological castles in the air which I had been building up for the last two years. One whole night I tried to think over the pleasure of seeing Shrewsbury again, but the barren plains of Peru gained the day. I made the following scheme (I know you will abuse me, and perhaps if I had put it in execution, my father would have sent a mandamus after me) ; it was to examine the Cordilleras of Chili during this summer, and in the winter go from port to port on the coast of Peru to Lima, returning this time next year to Valparaiso, cross the Cordilleras to Buenos Ayres, and take ship to England. Would not this have been a fine excursion, and in sixteen months I should have been with you all ? To have endured Tierra del Fuego and not seen the Pacific would have been miserable. . . .

I go on board to-morrow ; I have been for the last six weeks in Corfield's house. You cannot imagine what a kind friend I have found him. He is universally liked, and respected by the natives and foreigners. Several Chileno Signoritas are very obligingly anxious to become the signoras of this house. Tell my father I have kept my promise of being extravagant in Chili. I have drawn a bill of £100 (had it not better be notified to Messrs. Robarts & Co.) ; £50 goes to the Captain for the ensuing year, and £30 I take to sea for the small ports ; so that *bonâ fide* I have not spent £180 during these last four months. I hope not to draw another bill for six months. All the foregoing particulars were only settled yesterday. It has done me more good than a pint of medicine and I have not been so happy for the last year. If it had not been for my illness, these four months in Chili would have been very pleasant. I have had ill luck, however, in only one little earthquake having happened. I was lying in bed when there was a party at dinner in the house ; on a sudden I heard such a hubbub in the dining-room ; without a word being spoken, it was devil take the hindmost who should get out first ; at the same moment I felt my bed *slightly* vibrate in a lateral direction. The party were old stagers, and heard the noise which always precedes a shock ; and no old stager looks at an earthquake with philosophical eyes. . . .

Good-bye to you all ; you will not have another letter for some time.

My dear Catherine,
Yours affectionately,
CHAS. DARWIN.

My best love to my father, and all of you. Love to Nancy.

C. Darwin to Miss S. Darwin.

Valparaiso, April 23, 1835.

MY DEAR SUSAN,

I received, a few days since, your letter of November ; the three letters which I before mentioned are yet missing, but I do not doubt they will come to life. I returned a week ago from my excursion across the Andes to Mendoza. Since leaving England I have never made so successful a journey ; it has, however, been very expensive. I am sure my father would not regret it, if he could know how deeply I have enjoyed it : it was something more than enjoyment ; I cannot express the delight which I felt at such a famous winding-up of all my geology in South America. I literally could hardly sleep at nights for thinking over my day's

work. The scenery was so new, and so majestic ; everything at an elevation of 12,000 feet bears so different an aspect
from that in a lower country. I have seen many views more
beautiful, but none with so strongly marked a character.
To a geologist, also, there are such manifest proofs of
excessive violence ; the strata of the highest pinnacles are
tossed about like the crust of a broken pie.

I crossed by the Portillo Pass, which at this time of the
year is apt to be dangerous, so could not afford to delay
there. After staying a day in the stupid town of Mendoza, I
began my return by Uspallate, which I did very leisurely.
My whole trip only took up twenty-two days. I travelled
with, for me, uncommon comfort, as I carried a *bed!* My
party consisted of two Peons and ten mules, two of which
were with baggage, or rather food, in case of being snowed up.
Everything, however, favoured me ; not even a speck of this
year's snow had fallen on the road. I do not suppose any
of you can be much interested in geological details, but I
will just mention my principal results :—Besides understanding to a certain extent the description and manner of
the force which has elevated this great line of mountains,
I can clearly demonstrate that one part of the double line
is of an age long posterior to the other.   In the more ancient
line, which is the true chain of the Andes, I can describe the
sort and order of the rocks which compose it. These are
chiefly remarkable by containing a bed of gypsum nearly
2000 feet thick—a quantity of this substance I should think
unparalleled in the world.   What is of much greater consequence, I have procured fossil shells (from an elevation of
12,000 feet). I think an examination of these will give an
approximate age to these mountains, as compared to the
strata of Europe. In the other line of the Cordilleras there
is a strong presumption (in my own mind, conviction) that
the enormous mass of mountains, the peaks of which rise to
13,000 and 14,000 feet, are so very modern as to be con-

temporaneous with the plains of Patagonia (or about with the *upper* strata of the Isle of Wight). If this result shall be considered as proved,* it is a very important fact in the theory of the formation of the world ; because, if such wonderful changes have taken place so recently in the crust of the globe, there can be no reason for supposing former epochs of excessive violence. These modern strata are very remarkable by being threaded with metallic veins of silver, gold, copper, &c. ; hitherto these have been considered as appertaining to older formations. In these same beds, and close to a gold-mine, I found a clump of petrified trees, standing upright, with layers of fine sandstone deposited round them, bearing the impression of their bark. These trees are covered by other sandstones and streams of lava to the thickness of several thousand feet. These rocks have been deposited beneath water ; yet it is clear the spot where the trees grew must once have been above the level of the sea, so that it is certain the land must have been depressed by at least as many thousand feet as the superincumbent subaqueous deposits are thick. But I am afraid you will tell me I am prosy with my geological descriptions and theories. . . .

Your account of Erasmus' visit to Cambridge has made me long to be back there. I cannot fancy anything more delightful than his Sunday round of King's, Trinity, and those talking giants, Whewell and Sedgwick ; I hope your musical tastes continue in due force. I shall be ravenous for the pianoforte. . . .

I have not quite determined whether I will sleep at the ' Lion ' the first night when I arrive per ' Wonder,' or disturb you all in the dead of the night ; everything short of that is absolutely planned. Everything about Shrewsbury is growing in my mind bigger and more beautiful ; I am certain the

---

* The importance of these results has been fully recognized by geologists.

acacia and copper beech are two superb trees ; I shall know
every bush, and I will trouble you young ladies, when each of
you cut down your tree, to spare a few. As for the view
behind the house, I have seen nothing like it. It is the same
with North Wales ; Snowdon, to my mind, looks much
higher and much more beautiful than any peak in the Cordilleras. So you will say, with my benighted faculties, it is time
to return, and so it is, and I long to be with you. Whatever
the trees are, I know what I shall find all you. I am writing
nonsense, so farewell. My most affectionate love to all, and
I pray forgiveness from my father.

Yours most affectionately,

CHARLES DARWIN.

C. Darwin to W. D. Fox.

Lima, July, 1835.

MY DEAR FOX,

I have lately received two of your letters, one dated
June and the other November 1834 (they reached me, however,
in an inverted order). I was very glad to receive a history of
this most important year in your life. Previously I had only
heard the plain fact that you were married. You are a true
Christian and return good for evil, to send two such letters to
so bad a correspondent as I have been. God bless you for
writing so kindly and affectionately ; if it is a pleasure to
have friends in England, it is doubly so to think and know
that one is not forgotten, because absent. This voyage is
terribly long. I do so earnestly desire to return, yet I dare
hardly look forward to the future, for I do not know what
will become of me. Your situation is above envy : I do not
venture even to frame such happy visions. To a person fit to
take the office, the life of a clergyman is a type of all that is
respectable and happy. You tempt me by talking of your
fireside, whereas it is a sort of scene I never ought to think

about. I saw the other day a vessel sail for England ; it was quite dangerous to know how easily I might turn deserter. As for an English lady, I have almost forgotten what she is —something very angelic and good. As for the women in these countries, they wear caps and petticoats, and a very few have pretty faces, and then all is said. But if we are not wrecked on some unlucky reef, I will sit by that same fireside in Vale Cottage and tell some of the wonderful stories, which you seem to anticipate and, I presume, are not very ready to believe. *Gracias a dios,* the prospect of such times is rather shorter than formerly.

From this most wretched ' City of the Kings ' we sail in a fortnight, from thence to Guayaquil, Galapagos, Marquesas, Society Islands, &c., &c. I look forward to the Galapagos with more interest than any other part of the voyage. They abound with active volcanoes, and, I should hope, contain Tertiary strata. I am glad to hear you have some thoughts of beginning Geology. I hope you will ; there is so much larger a field for thought than in the other branches of Natural History. I am become a zealous disciple of Mr. Lyell's views, as known in his admirable book. Geologising in South America, I am tempted to carry parts to a greater extent even than he does. Geology is a capital science to begin, as it requires nothing but a little reading, thinking, and hammering. I have a considerable body of notes together ; but it is a constant subject of perplexity to me, whether they are of sufficient value for all the time I have spent about them, or whether animals would not have been of more certain value.

I shall indeed be glad once again to see you and tell you how grateful I feel for your steady friendship. God bless you, my very dear Fox.

Believe me,

Yours affectionately,

CHAS. DARWIN.

C. Darwin to J. S. Henslow.

Sydney, January, 1836.

MY DEAR HENSLOW,

This is the last opportunity of communicating with you before that joyful day when I shall reach Cambridge. I have very little to say : but I must write if it is only to express my joy that the last year is concluded, and that the present one, in which the *Beagle* will return, is gliding onwards. We have all been disappointed here in not finding even a single letter ; we are, indeed, rather before our expected time, otherwise I dare say, I should have seen your hand-writing. I must feed upon the future, and it is beyond bounds delightful to feel the certainty that within eight months I shall be residing once again most quietly in Cambridge. Certainly, I never was intended for a traveller ; my thoughts are always rambling over past or future scenes ; I cannot enjoy the present happiness for anticipating the future, which is about as foolish as the dog who dropped the real bone for its shadow.

*   *   *   *   *

In our passage across the Pacific we only touched at Tahiti and New Zealand; at neither of these places or at sea had I much opportunity of working. Tahiti is a most charming spot. Everything which former navigators have written is true. 'A new Cytheræa has risen from the ocean.' Delicious scenery, climate, manners of the people are all in harmony. It is, moreover, admirable to behold what the missionaries both here and at New Zealand have effected. I firmly believe they are good men working for the sake of a good cause. I much suspect that those who have abused or sneered at the missionaries, have generally been such as were not very anxious to find the natives moral and intelligent beings. During the remainder of our voyage we shall only visit places generally acknowledged as civilised,

and nearly all under the British flag. These will be a poor
field for Natural History, and without it I have lately discovered that the pleasure of seeing new places is as nothing.
I must return to my old resource and think of the future, but
that I may not become more prosy, I will say farewell till the
day arrives, when I shall see my Master in Natural History,
and can tell him how grateful I feel for his kindness and
friendship.

     Believe me, dear Henslow,
       Ever yours, most faithfully,
           CHAS. DARWIN.

     C. Darwin to Miss S. Darwin.

       Bahia, Brazil, August 4 [1836].
MY DEAR SUSAN,

  I will just write a few lines to explain the cause of
this letter being dated on the coast of South America. Some
singular disagreements in the longitudes made Captain Fitz-
Roy anxious to complete the circle in the southern hemisphere,
and then retrace our steps by our first line to England. This
zigzag manner of proceeding is very grievous; it has put the
finishing stroke to my feelings. I loathe, I abhor the sea and
all ships which sail on it. But I yet believe we shall reach
England in the latter half of October. At Ascension I
received Catherine's letter of October, and yours of November ;
the letter at the Cape was of a later date, but letters of all sorts
are inestimable treasures, and I thank you both for them.
The desert, volcanic rocks, and wild sea of Ascension, as soon
as I knew there was news from home, suddenly wore a
pleasing aspect, and I set to work with a good-will at my old
work of Geology. You would be surprised to know how
entirely the pleasure in arriving at a new place depends on
letters. We only stayed four days at Ascension, and then made
a very good passage to Bahia.

I little thought to have put my foot on South American coast again. It has been almost painful to find how much good enthusiasm has been evaporated during the last four years. I can now walk soberly through a Brazilian forest; not but what it is exquisitely beautiful, but now, instead of seeking for splendid contrasts, I compare the stately mango trees with the horse-chestnuts of England. Although this zigzag has lost us at least a fortnight, in some respects I am glad of it. I think I shall be able to carry away one vivid picture of inter-tropical scenery. We go from hence to the Cape de Verds ; that is, if the winds or the Equatorial calms will allow us. I have some faint hopes that a steady foul wind might induce the Captain to proceed direct to the Azores. For which most untoward event I heartily pray.

Both your letters were full of good news ; especially the expressions which you tell me Professor Sedgwick used about my collections. I confess they are deeply gratifying—I trust one part at least will turn out true, and that I shall act as I now think—as a man who dares to waste one hour of time has not discovered the value of life. Professor Sedgwick mentioning my name at all gives me hopes that he will assist me with his advice, of which, in my geological questions, I stand much in need. It is useless to tell you from the shameful state of this scribble that I am writing against time, having been out all morning, and now there are some strangers on board to whom I must go down and talk civility. Moreover, as this letter goes by a foreign ship, it is doubtful whether it will ever arrive. Farewell, my very dear Susan and all of you. Good-bye,

C. DARWIN.

C. Darwin to J. S. Henslow.

St. Helena, July 9, 1836.

MY DEAR HENSLOW,

I am going to ask you to do me a favour. I am very anxious to belong to the Geological Society. I do not know, but I suppose it is necessary to be proposed some time before being ballotted for; if such is the case, would you be good enough to take the proper preparatory steps? Professor Sedgwick very kindly offered to propose me before leaving England, if he should happen to be in London. I dare say he would yet do so.

I have very little to write about. We have neither seen, done, or heard of anything particular for a long time past; and indeed if at present the wonders of another planet could be displayed before us, I believe we should unanimously exclaim, what a consummate plague. No schoolboys ever sung the half sentimental and half jovial strain of 'dulce domum' with more fervour, than we all feel inclined to do. But the whole subject of 'dulce domum,' and the delight of seeing one's friends, is most dangerous, it must infallibly make one very prosy or very boisterous. Oh, the degree to which I long to be once again living quietly with not one single novel object near me! No one can imagine it till he has been whirled round the world during five long years in a ten-gun-brig. I am at present living in a small house (amongst the clouds) in the centre of the island, and within stone's throw of Napoleon's tomb. It is blowing a gale of wind with heavy rain and wretchedly cold; if Napoleon's ghost haunts his dreary place of confinement, this would be a most excellent night for such wandering spirits. If the weather chooses to permit me, I hope to see a little of the Geology (so often partially described) of the island. I suspect that differently from most volcanic islands its structure is rather complicated. It seems strange that this little centre of a distinct creation should, as is asserted, bear marks of recent elevation.

The *Beagle* proceeds from this place to Ascension, then to the Cape de Verds (what miserable places!) to the Azores to Plymouth, and then to home.   That most glorious of all days in my life will not, however, arrive till the middle of October. Some time in that month you will see me at Cambridge, where I must directly come to report myself to you, as my first Lord of the Admiralty.   At the Cape of Good Hope we all on board suffered a bitter disappointment in missing nine months' letters, which are chasing us from one side of the globe to the other. I dare say amongst them there was a letter from you; it is long since I have seen your hand-writing, but I shall soon see you yourself, which is far better.   As I am your pupil, you are bound to undertake the task of criticising and scolding me for all the things ill done and not done at all, which I fear I shall need much; but I hope for the best, and I am sure I have a good if not too easy taskmaster.

At the Cape Captain Fitz-Roy and myself enjoyed a memorable piece of good fortune in meeting Sir J. Herschel. We dined at his house and saw him a few times besides.   He was exceedingly good-natured, but his manners at first appeared to me rather awful.   He is living in a very comfortable country house, surrounded by fir and oak trees, which alone in so open a country, give a most charming air of seclusion and comfort.   He appears to find time for everything; he showed us a pretty garden full of Cape bulbs of his own collecting, and I afterwards understood that everything was the work of his own hands. . . . I am very stupid, and I have nothing more to say; the wind is whistling so mournfully over the bleak hills, that I shall go to bed and dream of England.

Good night, my dear Henslow,

Yours most truly obliged and affectionately,

CHAS. DARWIN.

C. Darwin to J. S. Henslow.

Shrewsbury, Thursday, October 6 [1836].

MY DEAR HENSLOW,

I am sure you will congratulate me on the delight of once again being home. The *Beagle* arrived at Falmouth on Sunday evening, and I reached Shrewsbury yesterday morning. I am exceedingly anxious to see you, and as it will be necessary in four or five days to return to London to get my goods and chattels out of the *Beagle*, it appears to me my best plan to pass through Cambridge. I want your advice on many points; indeed I am in the clouds, and neither know what to do or where to go. My chief puzzle is about the geological specimens—who will have the charity to help me in describing their mineralogical nature? Will you be kind enough to write to me one line by *return of post*, saying whether you are now at Cambridge? I am doubtful till I hear from Captain Fitz-Roy whether I shall not be obliged to start before the answer can arrive, but pray try the chance. My dear Henslow, I do long to see you ; you have been the kindest friend to me that ever man possessed. I can write no more, for I am giddy with joy and confusion.

Farewell for the present,

Yours most truly obliged,

CHARLES DARWIN.

C. Darwin to R. Fitz-Roy.

Shrewsbury, Thursday morning, October 6 [1836].

MY DEAR FITZ-ROY,

I arrived here yesterday morning at breakfast-time, and, thank God, found all my dear good sisters and father quite well. My father appears more cheerful and very little older than when I left. My sisters assure me I do not look the least different, and I am able to return the compliment.

Indeed, all England appears changed excepting the good old town of Shrewsbury and its inhabitants, which, for all I can see to the contrary, may go on as they now are to Doomsday. I wish with all my heart I was writing to you amongst your friends instead of at that horrid Plymouth. But the day will soon come, and you will be as happy as I now am. I do assure you I am a very great man at home ; the five years' voyage has certainly raised me a hundred per cent. I fear such greatness must experience a fall.

I am thoroughly ashamed of myself in what a dead-and-half-alive state I spent the few last days on board; my only excuse is that certainly I was not quite well. The first day in the mail tired me, but as I drew nearer to Shrewsbury everything looked more beautiful and cheerful. In passing Gloucestershire and Worcestershire I wished much for you to admire the fields, woods, and orchards. The stupid people on the coach did not seem to think the fields one bit greener than usual; but I am sure we should have thoroughly agreed that the wide world does not contain so happy a prospect as the rich cultivated land of England.

I hope you will not forget to send me a note telling me how you go on. I do indeed hope all your vexations and trouble with respect to our voyage, which we now know HAS an end, have come to a close. If you do not receive much satisfaction for all the mental and bodily energy you have expended in His Majesty's service, you will be most hardly treated. I put my radical sisters into an uproar at some of the prudent (if they were not honest Whigs, I would say shabby) proceedings of our Government. By the way, I must tell you for the honour and glory of the family that my father has a large engraving of King George IV. put up in his sitting-room. But I am no renegade, and by the time we meet my politics will be as firmly fixed and as wisely founded as ever they were.

I thought when I began this letter I would convince you

what a steady and sober frame of mind I was in.   But I find
I am writing most precious nonsense.   Two or three of our
labourers yesterday immediately set to work, and got most
excessively drunk in honour of the arrival of Master Charles.
Who then shall gainsay if Master Charles himself chooses
to make himself a fool.   Good-bye.   God bless you!   I hope
you are as happy, but much wiser, than your most sincere
but unworthy philosopher,

CHAS. DARWIN.

CHAPTER VII.

LONDON AND CAMBRIDGE.

1836–1842.

[THE period illustrated by the following letters includes the years between my father's return from the voyage of the *Beagle* and his settling at Down. It is marked by the gradual appearance of that weakness of health which ultimately forced him to leave London and take up his abode for the rest of his life in a quiet country house. In June 1841 he writes to Lyell: " My father scarcely seems to expect that I shall become strong for some years; it has been a bitter mortification for me to digest the conclusion that the 'race is for the strong,' and that I shall probably do little more, but be content to admire the strides others make in science."

There is no evidence of any intention of entering a profession after his return from the voyage, and early in 1840 he wrote to Fitz-Roy: " I have nothing .to wish for, excepting stronger health to go on with the subjects to which I have joyfully determined to devote my life."

These two conditions—permanent ill-health and a passionate love of scientific work for its own sake—determined thus early in his career, the character of his whole future life. They impelled him to lead a retired life of constant labour, carried on to the utmost limits of his physical power, a life which signally falsified his melancholy prophecy.

The end of the last chapter saw my father safely arrived at Shrewsbury on October 4, 1836, " after an absence of five years and two days." He wrote to Fox : "You cannot imagine how gloriously delightful my first visit was at home ; it was worth the banishment." But it was a pleasure that he could not long enjoy, for in the last days of October he was at Greenwich unpacking specimens from the *Beagle.* As to the destination of the collections he writes, somewhat despondingly, to Henslow :—

"I have not made much progress with the great men. I find, as you told me, that they are all overwhelmed with their own business. Mr. Lyell has entered, in the *most* good-natured manner, and almost without being asked, into all my plans. He tells me, however, the same story, that I must do all myself. Mr. Owen seems anxious to dissect some of the animals in spirits, and, besides these two, I have scarcely met any one who seems to wish to possess any of my specimens. I must except Dr. Grant, who is willing to examine some of the corallines. I see it is quite unreasonable to hope for a minute that any man will undertake the examination of a whole order. It is clear the collectors so much outnumber the real naturalists that the latter have no time to spare.

" I do not even find that the Collections care for receiving the unnamed specimens. The Zoological Museum * is nearly full, and upwards of a thousand specimens remain unmounted. I dare say the British Museum would receive them, but I cannot feel, from all I hear, any great respect even for the present state of that establishment. Your plan will be not only the best, but the only one, namely, to come down to Cambridge, arrange and group together the different families, and then wait till people, who are already working in different branches, may want specimens. But it appears to me [that] to do this it will be almost necessary to reside in London. As far as I can yet see my best plan will be to spend several

---

* The Museum of the Zoological   The collection was some years later
Society, then at 33 Bruton Street.   broken up and dispersed.

months in Cambridge, and then when, by your assistance, I
know on what ground I stand, to emigrate to London, where
I can complete my Geology and try to push on the Zoology.
I assure you I grieve to find how many things make me see
the necessity of living for some time in this dirty, odious
London. For even in Geology I suspect much assistance
and communication will be necessary in this quarter, for
instance, in fossil bones, of which none excepting the fragments of Megatherium have been looked at, and I clearly
see that without my presence they never would be. . . .

   " I only wish I had known the Botanists cared so much
for specimens * and the Zoologists so little ; the proportional
number of specimens in the two branches should have had
a very different appearance. I am out of patience with the
Zoologists, not because they are overworked, but for their
mean, quarrelsome spirit. I went the other evening to the
Zoological Society, where the speakers were snarling at each
other in a manner anything but like that of gentlemen.
Thank Heavens! as long as I remain in Cambridge there
will not be any danger of falling into any such contemptible
quarrels, whilst in London I do not see how it is to be
avoided. Of the Naturalists, F. Hope is out of London ;
Westwood I have not seen, so about my insects I know
nothing. I have seen Mr. Yarrell twice, but he is so evidently oppressed with business that it is too selfish to plague
him with my concerns. He has asked me to dine with the
Linnean on Tuesday, and on Wednesday I dine with the
Geological, so that I shall see all the great men. Mr. Bell,

---

   * A passage in a subsequent
letter shows that his plants also
gave him some anxiety. " I met
Mr. Brown a few days after you
had called on him ; he asked me in
rather an ominous manner what I
meant to do with my plants. In the
course of conversation Mr. Broderip, who was present, remarked
to him, ' You forget how long it is
since Captain King's expedition.'
He answered, ' Indeed, I have something in the shape of Captain King's
undescribed plants to make me
recollect it.' Could a better reason
be given, if I had been asked, by
me, for not giving the plants to the
British Museum ? "

I hear, is so much occupied that there is no chance of his wishing for specimens of reptiles. I have forgotten to mention Mr. Lonsdale,* who gave me a most cordial reception, and with whom I had much most interesting conversation. If I was not much more inclined for geology than the other branches of Natural History, I am sure Mr. Lyell's and Lonsdale's kindness ought to fix me. You cannot conceive anything more thoroughly good-natured than the heart-and-soul manner in which he put himself in my place and thought what would be best to do. At first he was all for London versus Cambridge, but at last I made him confess that, for some time at least, the latter would be for me much the best. There is not another soul whom I could ask, excepting yourself, to wade through and criticise some of those papers which I have left with you. Mr. Lyell owned that, second to London, there was no place in England so good for a Naturalist as Cambridge. Upon my word I am ashamed of writing so many foolish details; no young lady ever described her first ball with more particularity."

A few days later he writes more cheerfully : " I became acquainted with Mr. Bell,† who to my surprise expressed a good deal of interest about my crustacea and reptiles, and seems willing to work at them. I also heard that Mr. Broderip would be glad to look over the South American shells, so that things flourish well with me."

About his plants he writes with characteristic openness as to his own ignorance : " You have made me known amongst the botanists, but I felt very foolish when Mr. Don remarked

---

* William Lonsdale, b. 1794, d. 1871, was originally in the army, and served at the battles of Salamanca and Waterloo. After the war he left the service and gave himself up to science. He acted as assistant secretary to the Geological Society from 1829-42,

when he resigned, owing to ill-health.

† T. Bell, F.R.S., formerly Prof. of Zoology in King's College, London, and sometime secretary to the Royal Society. He afterwards described the reptiles for the zoology of the voyage of the *Beagle*.

on the beautiful appearance of some plant with an astounding
long name, and asked me about its habitation. Some one
else seemed quite surprised that I knew nothing about a Carex
from I do not know where. I was at last forced to plead
most entire innocence, and that I knew no more about the
plants which I had collected than the man in the moon."

As to part of his Geological Collection he was soon able
to write: "I [have] disposed of the most important part
[of] my collections, by giving all the fossil bones to the
College of Surgeons, casts of them will be distributed, and
descriptions published. They are very curious and valuable ;
one head belonged to some gnawing animal, but of the size
of a Hippopotamus! Another to an ant-eater of the size of a
horse!"

It is worth noting that at this time the only extinct mammalia from South America, which had been described, were
Mastodon (three species) and Megatherium. The remains
of the other extinct Edentata from Sir Woodbine Parish's
collection had not been described. My father's specimens
included (besides the above-mentioned Toxodon and Scelido-
therium) the remains of Mylodon, Glossotherium, another
gigantic animal allied to the ant-eater, and Macrauchenia.
His discovery of these remains is a matter of interest in itself,
but it has a special importance as a point in his own life,
since it was the vivid impression produced by excavating
them with his own hands * that formed one of the chief
starting-points of his speculations on the origin of species.
This is shown in the following extract from his Pocket
Book for this year (1837): "In July opened first note-book
on Transmutation of Species. Had been greatly struck from
about the month of previous March on character of South
American fossils, and species on Galapagos Archipelago.
These facts (especially latter), origin of all my views."]

* I have often heard him speak   a huge, partly excavated bone, when
of the despair with which he had to   the boat waiting for him would wait
break off the projecting extremity of   no longer.

1836-1837.

C. Darwin to W. D. Fox.

43 Great Marlborough Street,
November 6th [1836].

MY DEAR FOX,

I have taken a shamefully long time in answering your letter. But the busiest time of the whole voyage has been tranquillity itself to this last month. After paying Henslow a short but very pleasant visit, I came up to town to wait for the *Beagle's* arrival. At last I have removed all my property from on board, and sent the specimens of Natural History to Cambridge, so that I am now a free man. My London visit has been quite idle as far as Natural History goes, but has been passed in most exciting dissipation amongst the Dons in science. All my affairs, indeed, are most prosperous; I find there are plenty who will undertake the description of whole tribes of animals, of which I know nothing. So that about this day month I hope to set to work tooth and nail at the Geology, which I shall publish by itself.

It is quite ridiculous what an immensely long period it appears to me since landing at Falmouth. The fact is I have talked and laughed enough for years instead of weeks, so [that] my memory is quite confounded with the noise. I am delighted to hear you are turned geologist: when I pay the Isle of Wight a visit, which I am determined shall somehow come to pass, you will be a capital cicerone to the famous line of dislocation. I really suppose there are few parts of the world more interesting to a geologist than your island. Amongst the great scientific men, no one has been nearly so friendly and kind as Lyell. I have seen him several times, and feel inclined to like him much. You cannot imagine how good-naturedly he entered into all my plans. I speak now only of the London men, for Henslow was just like his former self, and therefore a most cordial and affectionate friend.

U 2

When you pay London a visit I shall be very proud to take you to the Geological Society, for be it known, I was proposed to be a F.G.S. last Tuesday. It is, however, a great pity that these and the other letters, especially F.R.S. are so very expensive.

I do not scruple to ask you to write to me in a week's time in Shrewsbury, for you are a good letter writer, and if people will have such good characters they must pay the penalty. Good-bye, dear Fox.

Yours,

C. D.

[His affairs being thus so far prosperously managed he was able to put into execution his plan of living at Cambridge, where he settled on December 10th, 1836. He was at first a guest in the comfortable home of the Henslows, but afterwards, for the sake of undisturbed work, he moved into lodgings. He thus writes to Fox, March 13th, 1837, from London :—

"My residence at Cambridge was rather longer than I expected, owing to a job which I determined to finish there, namely, looking over all my geological specimens. Cambridge yet continues a very pleasant, but not half so merry a place as before. To walk through the courts of Christ's College, and not know an inhabitant of a single room, gave one a feeling half melancholy. The only evil I found in Cambridge was its being too pleasant : there was some agreeable party or another every evening, and one cannot say one is engaged with so much impunity there as in this great city."

A trifling record of my father's presence in Cambridge occurs in the book kept in Christ's College combination-room, where fines and bets were recorded, the earlier entries giving a curious impression of the after-dinner frame of mind of the fellows. The bets were not allowed to be made in money, but were, like the fines, paid in wine. The bet which my father made and lost is thus recorded :—

"*Feb.* 23, 1837.—Mr. Darwin *v.* Mr. Baines, that the combination-room measures from the ceiling to the floor more
than (x) feet.   1 Bottle paid same day.
"N.B. Mr. Darwin may measure at any part of the room he
pleases."

Besides arranging the geological and mineralogical specimens, he had his 'Journal of Researches' to work at, which
occupied his evenings at Cambridge. He also read a short
paper at the Zoological Society,* and another at the Geological
Society,† on the recent elevation of the coast of Chili.

Early in the spring of 1837 (March 6th) he left Cambridge
for London, and a week later he was settled in lodgings at
36 Great Marlborough Street; and except for a "short visit to
Shrewsbury" in June, he worked on till September, being
almost entirely employed on his 'Journal.' He found time,
however, for two papers at the Geological Society.‡

He writes of his work to Fox (March, 1837):—

"In your last letter you urge me to get ready *the* book.
I am now hard at work and give up everything else for it.
Our plan is as follows: Capt. Fitz-Roy writes two volumes
out of the materials collected during the last voyage under
Capt. King to Tierra del Fuego, and during our circumnavigation. I am to have the third volume, in which I intend
giving a kind of journal of a naturalist, not following, however, always the order of time, but rather the order of position. The habits of animals will occupy a large portion,
sketches of the geology, the appearance of the country, and
personal details will make the hodge-podge complete. Afterwards I shall write an account of the geology in detail, and

* "Notes upon Rhea Americana,"
'Zool. Soc. Proc.' v. 1837, pp. 35, 36.
  † 'Geol. Soc. Proc.' ii. 1838, pp.
446–449.
  ‡ "A sketch of the deposits containing extinct mammalia in the
neighbourhood of the Plata," 'Geol.

Soc. Proc.' ii. 1838, pp. 542–544;
and "On certain areas of elevation
and subsidence in the Pacific and
Indian oceans, as deduced from
the study of coral formations,"
'Geol. Soc. Proc.' ii. 1838, pp. 552–
554.

draw up some zoological papers. So that I have plenty of work for the next year or two, and till that is finished I will have no holidays."

Another letter to Fox (July) gives an account of the progress of his work :—

"I gave myself a holiday and a visit to Shrewsbury [in June], as I had finished my Journal. I shall now be very busy in filling up gaps and getting it quite ready for the press by the first of August. I shall always feel respect for every one who has written a book, let it be what it may, for I had no idea of the trouble which trying to write common English could cost one. And, alas, there yet remains the worst part of all, correcting the press. As soon as ever that is done I must put my shoulder to the wheel and commence at the Geology. I have read some short papers to the Geological Society, and they were favourably received by the great guns, and this gives me much confidence, and I hope not a very great deal of vanity, though I confess I feel too often like a peacock admiring his tail. I never expected that my Geology would ever have been worth the consideration of such men as Lyell, who has been to me, since my return, a most active friend. My life is a very busy one at present, and I hope may ever remain so ; though Heaven knows there are many serious drawbacks to such a life, and chief amongst them is the little time it allows one for seeing one's natural friends. For the last three years, I have been longing and longing to be living at Shrewsbury, and after all now in the course of several months, I see my good dear people at Shrewsbury for a week. Susan and Catherine have, however, been staying with my brother here for some weeks, but they had returned home before my visit."

Besides the work already mentioned he had much to busy him in making arrangements for the publication of the 'Zoology of the Voyage of the *Beagle.*' The following letters illustrate this subject.]

C. Darwin to L. Jenyns.[*]

36 Great Marlborough Street,
April 10th, 1837.

DEAR JENYNS,

During the last week several of the zoologists of this place have been urging me to consider the possibility of publishing the ' Zoology of the *Beagle's* Voyage' on some uniform plan. Mr. Macleay† has taken a great deal of interest in the subject, and maintains that such a publication is very desirable, because it keeps together a series of observations made respecting animals inhabiting the same part of the world, and allows any future traveller taking them with him. How far this facility of reference is of any consequence I am very doubtful ; but if such is the case, it would be more satisfactory to myself to see the gleanings of my hands, after having passed through the brains of other naturalists, collected together in one work.   But such considerations ought not to have much weight.   The whole scheme is at present merely floating in the air ; but I was determined to let you know, as I should much like to know what you think about it, and whether you would object to supply descriptions of the fish to such a work instead of to ' Transactions.'   I apprehend the whole will be impracticable, without Government will aid in engraving the plates, and this I fear is a mere chance, only I think I can put in a strong claim, and get myself well backed by the naturalists of this place, who nearly all take a good deal

---

* Now Rev. L. Blomefield.

† William Sharp Macleay was the son of Alexander Macleay, formerly Colonial Secretary of New South Wales, and for many years Secretary of the Linnean Society. The son, who was a most zealous Naturalist, and had inherited from his father a very large general collection of insects, made Entomology his chief study, and gained great notoriety by his now forgotten *Quinary System*, set forth in the Second Part of his ' Horæ Entomologicæ,' published in 1821.—[I am indebted to Rev. L. Blomefield for the foregoing note.]

of interest in my collections. I mean to-morrow to see
Mr. Yarrell; if he approves, I shall begin and take more
active steps; for I hear he is most prudent and most wise.
It is scarcely any use speculating about any plan, but I
thought of getting subscribers and publishing the work in
parts (as long as funds would last, for I myself will not lose
money by it). In such case, whoever had his own part ready
on any order might publish it separately (and ultimately the
parts might be sold separately), so that no one should be
delayed by the other. The plan would resemble, on a humble
scale, Ruppel's 'Atlas,' or Humboldt's 'Zoologie,' where
Latreille, Cuvier, &c., wrote different parts. I myself should
have little to do with it; excepting in some orders adding
habits and ranges, &c., and geographical sketches, and perhaps
afterwards some descriptions of invertebrate animals . . . . .

I am working at my Journal; it gets on slowly, though
I am not idle. I thought Cambridge a bad place from good
dinners and other temptations, but I find London no better,
and I fear it may grow worse. I have a capital friend in
Lyell, and see a great deal of him, which is very advantageous
to me in discussing much South American geology. I miss
a walk in the country very much; this London is a vile smoky
place, where a man loses a great part of the best enjoyments
in life. But I see no chance of escaping, even for a week,
from this prison for a long time to come. I fear it will be
some time before we shall meet; for I suppose you will not
come up here during the spring, and I do not think I shall be
able to go down to Cambridge. How I should like to have a
good walk along the Newmarket road to-morrow, but Oxford
Street must do instead. I do hate the streets of London. Will
you tell Henslow to be careful with the *edible* fungi from Tierra
del Fuego, for I shall want some specimens for Mr. Brown,
who seems *particularly* interested about them. Tell Henslow,
I think my silicified wood has unflintified Mr. Brown's heart,
for he was very gracious to me, and talked about the Gala-

pagos plants; but before he never would say a word. It is just striking twelve o'clock; so I will wish you a very good night.

My dear Jenyns,
Yours most truly,
C. DARWIN.

[A few weeks later the plan seems to have been matured, and the idea of seeking Government aid to have been adopted.]

C. Darwin to J. S. Henslow.

36 Great Marlborough Street,
[18th May, 1837].

MY DEAR HENSLOW,

I was very glad to receive your letter. I wanted much to hear how you were getting on with your manifold labours. Indeed I do not wonder your head began to ache; it is almost a wonder you have any head left. Your account of the Gamlingay expedition was cruelly tempting, but I cannot anyhow leave London. I wanted to pay my good, dear people at Shrewsbury a visit of a few days, but I found I could not manage it; at present I am waiting for the signatures of the Duke of Somerset, as President of the Linnean, and of Lord Derby and Whewell, to a statement of the value of my collection; the instant I get this I shall apply to Government for assistance in engraving, and so publish the 'Zoology' on some uniform plan. It is quite ridiculous the time any operation requires which depends on many people.

I have been working very steadily, but have only got two-thirds through the Journal part alone. I find, though I remain daily many hours at work, the progress is very slow: it is an awful thing to say to oneself, every fool and every clever man in England, if he chooses, may make as many ill-natured remarks as he likes on this unfortunate sentence.

*   *   *   *   *

[In August he writes to Henslow to announce the success of the scheme for the publication of the ' Zoology of the Voyage of the *Beagle*,' through the promise of a grant of £1000 from the Treasury : " I have delayed writing to you, to thank you most sincerely for having so effectually managed my affair. I waited till I had an interview with the Chancellor of the Exchequer.* He appointed to see me this morning, and I had a long conversation with him, Mr. Peacock being present. Nothing could be more thoroughly obliging and kind than his whole manner. He made no sort of restriction, but only told me to make the most of [the] money, which of course I am right willing to do.

" I expected rather an awful interview, but I never found anything less so in my life. It will be my fault if I do not make a good work ; but I sometimes take an awful fright that I have not materials enough. It will be excessively satisfactory at the end of some two years to find all materials made the most they were capable of."

Later in the autumn he wrote to Henslow : " I have not been very well of late, with an uncomfortable palpitation of the heart, and my doctors urge me *strongly* to knock off all work, and go and live in the country for a few weeks." He accordingly took a holiday of about a month at Shrewsbury and Maer, and paid Fox a visit in the Isle of Wight. It was, I believe, during this visit, at Mr. Wedgwood's house at Maer, that he made his -first observations on the work done by earthworms, and late in the autumn he read a paper on the subject at the Geological Society.† During these two months he was also busy preparing the scheme of the ' Zoology of the Voyage of the *Beagle*,' and in beginning to put together the Geological results of his travels.

The following letter refers to the proposal that he should take the Secretaryship of the Geological Society.]

* T. Spring Rice.　　　　　' Geol. Soc. Proc.' ii. 1838, pp. 574–
† " On the formation of mould,"　576.

C. Darwin to J. S. Henslow.

October 14th [1837].

MY DEAR HENSLOW,

... I am much obliged to you for your message about the Secretaryship. I am exceedingly anxious for you to hear my side of the question, and will you be so kind as afterwards to give me your fair judgment. The subject has haunted me all summer. I am unwilling to undertake the office for the following reasons : First, my entire ignorance of English Geology, a knowledge of which would be almost necessary in order to shorten many of the papers before reading them before the Society, or rather to know what parts to skip. Again, my ignorance of all languages, and not knowing how to pronounce even a *single* word of French— a language so perpetually quoted. It would be disgraceful to the Society to have a Secretary who could not read French. Secondly, the loss of time ; pray consider that I should have to look after the artists, superintend and furnish materials for the Government work, which will come out in parts, and which must appear regularly. All my Geological notes are in a very rough state ; none of my fossil shells worked up ; and I have much to read. I have had hopes, by giving up society and not wasting an hour, that I should finish my Geology in a year and a half, by which time the description of the higher animals by others would be completed, and my whole time would then necessarily be required to complete myself the description of the invertebrate ones. If this plan fails, as the Government work must go on, the Geology would necessarily be deferred till probably at least three years from this time. In the present state of the science, a great part of the utility of the little I have done would be lost, and all freshness and pleasure quite taken from me.

I know from experience the time required to make abstracts

even of my own papers for the 'Proceedings.' If I was
Secretary, and had to make double abstracts of each paper,
studying them before reading, and attendance would *at least*
cost me three days (and often more) in the fortnight.   There
are likewise other accidental and contingent losses of time ;
I know Dr. Royle found the office consumed much of his
time.   If by merely giving up any amusement, or by working
harder than I have done, I could save time, I would undertake
the Secretaryship ; but I appeal to you whether, with my
slow manner of writing, with two works in hand, and with
the certainty, if I cannot complete the Geological part within
a fixed period, that its publication must be retarded for a
very long time,—whether any Society whatever has any
claim on me for three days' disagreeable work every fortnight.
I cannot agree that it is a duty on my part, as a follower
of science, as long as I devote myself to the completion of
the work I have in hand, to delay that, by undertaking what
may be done by any person who happens to have more
spare time than I have at present.   Moreover, so early in
my scientific life, with so very much as I have to learn,
the office, though no doubt a great honour, &c., for me,
would be the more burdensome.   Mr. Whewell (I know very
well), judging from himself, will think I exaggerate the time
the Secretaryship would require ; but I absolutely know the
time which with me the simplest writing consumes.   I do
not at all like appearing so selfish as to refuse Mr. Whewell,
more especially as he has always shown, in the kindest
manner, an interest in my affairs.   But I cannot look forward with even tolerable comfort to undertaking an office
without entering on it heart and soul, and that would be
impossible with the Government work and the Geology in
hand.

My last objection is, that I doubt how far my health will
stand the confinement of what I have to do, without any
additional work.   I merely repeat, that you may know I am

not speaking idly, that when I consulted Dr. Clark in town, he at first urged me to give up entirely all writing and even correcting press for some weeks. Of late anything which flurries me completely knocks me up afterwards, and brings on a violent palpitation of the heart. Now the Secretaryship would be a periodical source of more annoying trouble to me than all the rest of the fortnight put together. In fact, till I return to town, and see how I get on, if I wished the office ever so much, I *could* not say I would positively undertake it. I beg of you to excuse this very long prose all about myself, but the point is one of great interest. I can neither bear to think myself very selfish and sulky, nor can I see the possibility of my taking the Secretaryship without making a sacrifice of all my plans and a good deal of comfort.

If you see Whewell, would you tell him the substance of this letter ; or, if he will take the trouble, he may read it. My dear Henslow, I appeal to you *in loco parentis.* Pray tell me what you think? But do not judge me by the activity of mind which you and a few others possess, for in that case the more different things in hand the pleasanter the work ; but, though I hope I never shall be idle, such is not the case with me.

Ever, dear Henslow,
Yours most truly,
C. DARWIN.

[He ultimately accepted the post, and held it for three years —from February 16, 1838, to February 19, 1841.

After being assured of the Grant for the publication of the Zoology of the Voyage of the *Beagle*,' there was much to be done in arranging the scheme of publication, and this occupied him during part of October and November.]

C. Darwin to J. S. Henslow.

[4th November, 1837.]

MY DEAR HENSLOW,

. . . Pray tell Leonard * that my Government work is going on smoothly, and I hope will be prosperous. He will see in the Prospectus his name attached to the fish; I set my shoulders to the work with a good heart. I am very much better than I was during the last month before my Shrewsbury visit. I fear the Geology will take me a great deal of time; I was looking over one set of notes, and the quantity I found I had to read, for that one place was frightful. If I live till I am eighty years old I shall not cease to marvel at finding myself an author; in the summer before I started, if any one had told me that I should have been an angel by this time, I should have thought it an equal impossibility. This marvellous transformation is all owing to you.

I am sorry to find that a good many errata are left in the part of my volume, which is printed. During my absence Mr. Colburn employed some goose to revise, and he has multiplied, instead of diminishing my oversights: but for all that, the smooth paper and clear type has a charming appearance, and I sat the other evening gazing in silent admiration at the first page of my own volume, when I received it from the printers!

Good bye, my dear Henslow,

C. DARWIN.

1838.

[From the beginning of this year to nearly the end of June, he was busily employed on the zoological and geological results of his voyage. This spell of work was interrupted

* Rev. L. Jenyns.

only by a visit of three days to Cambridge, in May ; and even this short holiday was taken in consequence of failing health, as we may assume from the entry in his diary : "May 1st, unwell," and from a letter to his sister (May 16, 1838), when he wrote :—

"My trip of three days to Cambridge has done me such wonderful good, and filled my limbs with such elasticity, that I must get a little work out of my body before another holiday." This holiday seems to have been thoroughly enjoyed ; he wrote to his sister :—

"Now for Cambridge : I stayed at Henslow's house and enjoyed my visit extremely. My friends gave me a most cordial welcome. Indeed, I was quite a lion there. Mrs. Henslow unfortunately was obliged to go on Friday for a visit in the country. That evening we had at Henslow's a brilliant party of all the geniuses in Cambridge, and a most remarkable set of men they most assuredly are. On Saturday I rode over to L. Jenyns', and spent the morning with him. I found him very cheerful, but bitterly complaining of his solitude. On Saturday evening dined at one of the Colleges, played at bowls on the College Green after dinner, and was deafened with nightingales singing. Sunday, dined in Trinity ; capital dinner, and was very glad to sit by Professor Lee* . . . ; I find him a very pleasant chatting man, and in high spirits like a boy, at having lately returned from a living or a curacy, for seven years in Somersetshire, to civilised society and oriental manuscripts. He had exchanged his living to one within fourteen miles of Cambridge, and seemed perfectly happy. In the evening attended Trinity Chapel, and heard 'The Heavens are telling the Glory of God,' in magnificent style ; the last chorus seemed to shake the very walls of the College. After chapel a large party in Sedgwick's rooms. So much for my Annals."

* Samuel Lee, of Queens', was Professor of Arabic from 1819 to 1831, and Regius Professor of Hebrew from 1831 to 1848.

He started, towards the end of June, on his expedition to Glen Roy, of which he writes to Fox: "I have not been very well of late, which has suddenly determined me to leave London earlier than I had anticipated. I go by the steam-packet to Edinburgh,—take a solitary walk on Salisbury Craigs, and call up old thoughts of former times, then go on to Glasgow and the great valley of Inverness, near which I intend stopping a week to geologise the parallel roads of Glen Roy, thence to Shrewsbury, Maer for one day, and London for smoke, ill-health and hard work."

He spent "eight good days" over the Parallel Roads. His Essay on this subject was written out during the same summer, and published by the Royal Society.* He wrote in his Pocket Book: "September 6 [1838]. Finished the paper on 'Glen Roy,' one of the most difficult and instructive tasks I was ever engaged on." It will be remembered that in his 'Recollections' he speaks of this paper as a failure, of which he was ashamed.

At the time at which he wrote, the latest theory of the formation of the Parallel Roads was that of Sir Lauder Dick and Dr. Macculloch, who believed that lakes had anciently existed in Glen Roy, caused by dams of rock or alluvium. In arguing against this theory he conceived that he had dis-proved the admissibility of any lake theory, but in this point he was mistaken. He wrote (Glen Roy paper, p. 49) "the conclusion is inevitable, that no hypothesis founded on the supposed existence of a sheet of water confined by *barriers*, that is a lake, can be admitted as solving the problematical origin of the parallel roads of Lochaber."

Mr. Archibald Geikie has been so good as to allow me to quote a passage from a letter addressed to me (Nov. 19, 1884) in compliance with my request for his opinion on the character of my father's Glen Roy work :—

"Mr. Darwin's 'Glen Roy' paper, I need not say, is marked

* 'Phil. Trans.' 1839, pp. 39–82.

by all his characteristic acuteness of observation and determination to consider all possible objections. It is a curious example, however, of the danger of reasoning by a method of exclusion in Natural Science. Finding that the waters which formed the terraces in the Glen Roy region could not possibly have been dammed back by barriers of rock or of detritus, he saw no alternative but to regard them as the work of the sea. Had the idea of transient barriers of glacier-ice occurred to him, he would have found the difficulties vanish from the lake-theory which he opposed, and he would not have been unconsciously led to minimise the altogether overwhelming objections to the supposition that the terraces are of marine origin."

It may be added that the idea of the barriers being formed by glaciers could hardly have occurred to him, considering what was the state of knowledge at the time, and bearing in mind what his want of opportunities of observing glacial action on a large scale.

The latter half of July was passed at Shrewsbury and Maer. The only entry of any interest is one of being "very idle" at Shrewsbury, and of opening "a note-book connected with metaphysical inquiries." In August he records that he read "a good deal of various amusing books, and paid some attention to metaphysical subjects."

The work done during the remainder of the year comprises the book on coral reefs (begun in October), and some work on the phenomena of elevation in S. America.]

C. Darwin to C. Lyell.

36 Great Marlborough Street,
August 9th [1838].

MY DEAR LYELL,

I did not write to you at Norwich, for I thought I should have more to say, if I waited a few more days. Very many thanks for the present of your 'Elements,' which I

received (and I believe the *very first* copy distributed) together
with your note. I have read it through every word, and am
full of admiration of it, and, as I now see no geologist, I must
talk to you about it. There is no pleasure in reading a book
if one cannot have a good talk over it ; I repeat, I am full of
admiration of it, it is as clear as daylight, in fact I felt in
many parts some mortification at thinking how geologists
have laboured and struggled at proving what seems, as you
have put it, so evidently probable. I read with much interest
your sketch of the secondary deposits ; you have contrived to
make it quite "juicy," as we used to say as children of a good
story. There was also much new to me, and I have to copy
out some fifty notes and references. It must do good, the
heretics against common sense must yield. . . . By the way,
do you recollect my telling you how much I disliked the
manner ———— referred to his other works, as much as to say,
"You must, ought, and shall buy everything I have written."
To my mind, you have somehow quite avoided this ; your
references only seem to say, " I can't tell you all in this work,
else I would, so you must go to the 'Principles' ; and many a
one, I trust, you will send there, and make them, like me,
adorers of the good science of rock-breaking. You will see I
am in a fit of enthusiasm, and good cause I have to be, when
I find you have made such infinitely more use of my Journal
than I could have anticipated. I will say no more about the
book, for it is all praise. I must, however, admire the elaborate honesty with which you quote the words of all living and
dead geologists.

My Scotch expedition answered brilliantly ; my trip in the
steam-packet was absolutely pleasant, and I enjoyed the spectacle, wretch that I am, of two ladies, and some small children
quite sea-sick, I being well. Moreover, on my return from
Glasgow to Liverpool, I triumphed in a similar manner over
some full-grown men. I stayed one whole day in Edinburgh,
or more truly on Salisbury Craigs ; I want to hear some day

what you think about that classical ground,—the structure
was to me new and rather curious,—that is, if I understand it
right. I crossed from Edinburgh in gigs and carts (and carts
without springs, as I never shall forget) to Loch Leven. I
was disappointed in the scenery, and reached Glen Roy on
Saturday evening, one week after leaving Marlborough Street.
Here I enjoyed five [?] days of the most beautiful weather with
gorgeous sunsets, and all nature looking as happy as I felt.
I wandered over the mountains in all directions, and examined
that most extraordinary district. I think, without any exceptions, not even the first volcanic island, the first elevated
beach, or the passage of the Cordillera, was so interesting to
me as this week. It is far the most remarkable area I ever
examined. I have fully convinced myself (after some doubting at first) that the shelves are sea-beaches, although I could
not find a trace of a shell; and I think I can explain away
most, if not all, the difficulties. I found a piece of a road in
another valley, not hitherto observed, which is important;
and I have some curious facts about erratic blocks, one of
which was perched up on a peak 2200 feet above the sea. I
am now employed in writing a paper on the subject, which I
find very amusing work, excepting that I cannot anyhow con-
dense it into reasonable limits. At some future day I hope
to talk over some of the conclusions with you, which the
examination of Glen Roy has led me to. Now I have had
my talk out, I am much easier, for I can assure you Glen Roy
has astonished me.

I am living very quietly, and therefore pleasantly, and am
crawling on slowly but steadily with my work. I have come
to one conclusion, which you will think proves me to be
a very sensible man, namely, that whatever you say proves
right; and as a proof of this, I am coming into your way of
only working about two hours at a spell; I then go out and
do my business in the streets, return and set to work again,
and thus make two separate days out of one. The new plan

X 2

answers capitally ; after the second half day is finished I go
and dine at the Athenæum like a gentleman, or rather like a
lord, for I am sure the first evening I sat in that great drawing-room, all on a sofa by myself, I felt just like a duke.   I
am full of admiration at the Athenæum, one meets so many
people there that one likes to see.   The very first time I
dined there (*i.e.* last week) I met Dr. Fitton * at the door, and
he got together quite a party—Robert Brown, who is gone to
Paris and Auvergne, Macleay [?] and Dr. Boott.†   Your helping
me into the Athenæum has not been thrown away, and I
enjoy it the more because I fully expected to detest it.

I am writing you a most unmerciful letter, but I shall get
Owen to take it to Newcastle.   If you have a mind to be a
very generous man you will write to me from Kinnordy,‡ and
tell me some Newcastle news, as well as about the Craig, and
about yourself and Mrs. Lyell, and everything else in the
world.   I will send by Hall the ' Entomological Transactions,'
which I have borrowed for you ; you will be disappointed in
——'s papers, that is if you suppose my dear friend has a
single clear idea upon any one subject.   He has so involved
recent insects and true fossil insects in one table that I fear
you will not make much out of it, though it is a subject which
ought I should think to come into the ' Principles.'   You will

* W. H. Fitton (b. 1780, d. 1861) was a physician and geologist, and sometime president of the Geological Society.   He established the ' Proceedings,' a mode of publication afterwards adopted by other societies.

† Francis Boott (b. 1792, d. 1863) is chiefly known as a botanist through his work on the genus Carex.   He was also well known in connection with the Linnean Society of which he was for many years an office-bearer.   He is described (in a biographical sketch published in the *Gardeners' Chronicle*, 1864) as having been one of the first physicians in London who gave up the customary black coat, knee-breeches and silk stockings, and adopted the ordinary dress of the period, a blue coat with brass buttons, and a buff waistcoat, a costume which he continued to wear to the last.   After giving up practice, which he did early in life, he spent much of his time in acts of unpretending philanthropy.

‡ The house of Lyell's father.

be amused at some of the ridiculo-sublime passages in the
papers, and no doubt will feel acutely a sneer there is at yourself.   I have heard from more than one quarter that quarrelling is expected at Newcastle * ; I am sorry to hear it.   I met
old —— this evening at the Athenæum, and he muttered
something about writing to you or some one on the subject ;
I am however all in the dark.   I suppose, however, I shall be
illuminated, for I am going to dine with him in a few days, as
my inventive powers failed in making any excuse.   A friend
of mine dined with him the other day, a party of four, and
they finished ten bottles of wine—a pleasant prospect for me ;
but I am determined not even to taste his wine, partly for the
fun of seeing his infinite disgust and surprise. . . .

I pity you the infliction of this most unmerciful letter.
Pray remember me most kindly to Mrs. Lyell when you arrive
at Kinnordy.   I saw her name in the landlord's book of In-
verorum.   Tell Mrs. Lyell to read the second series of 'Mr.
Slick of Slickville's Sayings.' . . . He almost beats "Samivel,"
that prince of heroes.   Good night, my dear Lyell ; you will
think I have been drinking some strong drink to write so
much nonsense, but I did not even taste Minerva's small beer
to-day.

Yours most sincerely,
CHAS. DARWIN.

C. Darwin to C. Lyell.

Friday night, September 13th [1838].

MY DEAR LYELL,

I was astonished and delighted at your gloriously
long letter, and I am sure I am very much obliged to
Mrs. Lyell for having taken the trouble to write so much.†
I mean to have a good hour's enjoyment and scribble away

* At the meeting of the British
Association.

† Lyell dictated much of his
correspondence.

to you, who have so much geological sympathy that I do not care how egotistically I write. . . .

I have got so much to say about all sorts of trifling things that I hardly know what to begin about. ˙I need not say how pleased I am to hear that Mr. Lyell * likes my Journal. To hear such tidings is a kind of resurrection, for I feel towards my first-born child as if it had long since been dead, buried, and forgotten ; but the past is nothing and the future everything to us geologists, as you show in your capital motto to the 'Elements.' By the way, have you read the article, in the 'Edinburgh Review,' on M. Comte, 'Cours de la Philosophie ' (or some such title) ? It is capital ; there are some fine sentences about the very essence of science being prediction, which reminded me of " its law being progress."

I will now begin and go through your letter *seriatim.* I dare say your plan of putting the Elie de Beaumont's chapter separately and early will be very good ; anyhow, it is showing a bold front in the first edition which is to be translated into French. It will be a curious point to geologists hereafter to note how long a man's name will support a theory so completely exposed as that of De Beaumont's has been by you ; you say you " begin to hope that the great principles there insisted on will stand the test of time." *Begin to hope :* why, the *possibility* of a doubt has never crossed my mind for many a day. This may be very un-philosophical, but my geological salvation is staked on it. After having just come back from Glen Roy, and found how difficulties smooth away under your principles, it makes me quite indignant that you should talk of *hoping.* With respect to the question, how far my coral theory bears on De Beaumont's theory, I think it would be prudent to quote me with great caution until my whole account is published, and then you (and others) can judge how far there is founda-tion for such generalisation. Mind, I do not doubt its truth ;

* Father of the geologist.

but the extension of any view over such large spaces, from comparatively few facts, must be received with much caution. I do not myself the least doubt that within the recent (or as you, much to my annoyment, would call it, "New Pliocene") period, tortuous bands—not all the bands parallel to each other—have been elevated and corresponding ones subsided, though within the same period some parts probably remained for a time stationary, or even subsided. I do not believe a more utterly false view could have been invented than great straight lines being suddenly thrown up.

When my book on Volcanoes and Coral Reefs will be published I hardly know ; I fear it will be at least four or five months ; though, mind, the greater part is written. I find so much time is lost in correcting details and ascertaining their accuracy. The Government Zoological work is a millstone round my neck, and the Glen Roy paper has lost me six weeks. I will not, however, say lost ; for, supposing I can prove to others' satisfaction what I have convinced myself is the case, the inference I think you will allow to be important. I cannot doubt that the molten matter beneath the earth's crust possesses a high degree of fluidity, almost like the sea beneath the block ice. By the way, I hope you will give me some Swedish case to quote, of shells being preserved on the surface, but not in contemporaneous beds of gravel. . . .

Remember what I have often heard you say : the country is very bad for the intellects ; the Scotch mists will put out some volcanic speculations. You see I am affecting to become very Cockneyfied, and to despise the poor country-folk, who breathe fresh air instead of smoke, and see the goodly fields instead of the brick houses in Marlborough Street, the very sight of which I confess I abhor. I am glad to hear what a favourable report you give of the British Association. I am the more pleased because I have been fighting its battle with Basil Hall, Stokes, and several others, having made up my mind, from the report in the *Athenæum*, that it must have been an excellent meeting. I have been much amused with an account I have received of the wars of Don Roderick * and Babbage. What a grievous pity it is that the latter should be so implacable . . . This is a most rigmarole letter, for after each sentence I take breath, and you will have need of it in reading it. . . .

I wish with all my heart that my Geological book was out. I have every motive to work hard, and will, following your steps, work just that degree of hardness to keep well. I should like my volume to be out before your new edition of 'Principles' appears. Besides the Coral theory, the volcanic chapters will, I think, contain some new facts. I have lately been sadly tempted to be idle—that is, as far as pure geology is concerned—by the delightful number of new views which have been coming in thickly and steadily,—on the classification and affinities and instincts of animals—bearing on the question of species. Note-book after note-book has been filled with facts which begin to group themselves *clearly* under sub-laws.

Good night, my dear Lyell. I have filled my letter and enjoyed my talk to you as much as I can without having you *in propriâ personâ*. Think of the bad effects of the country—so once more good night.

Ever yours,
CHAS. DARWIN.

Pray again give my best thanks to Mrs. Lyell.

[The record of what he wrote during the year does not give a true index of the most important work that was in progress,—the laying of the foundation-stones of what was to be the achievement of his life. This is shown in the foregoing letter to Lyell, where he speaks of being "idle," and the following extract from a letter to Fox, written in June, is of interest in this point of view:

* Murchison.

"I am delighted to hear you are such a good man as not to have forgotten my questions about the crossing of animals. It is my prime hobby, and I really think some day I shall be able to do something in that most intricate subject, species and varieties."]

1839 to 1841.

[In the winter of 1839 (Jan. 29) my father was married to his cousin, Emma Wedgwood.* The house in which they lived for the first few years of their married life, No. 12 Upper Gower Street, was a small common-place London house, with a drawing-room in front, and a small room behind, in which they lived for the sake of quietness. In later years my father used to laugh over the surpassing ugliness of the furniture, carpets, &c., of the Gower Street house. The only redeeming feature was a better garden than most London houses have, a strip as wide as the house, and thirty yards long. Even this small space of dingy grass made their London house more tolerable to its two country-bred inhabitants.

Of his life in London he writes to Fox (October 1839): "We are living a life of extreme quietness; Delamere itself, which you describe as so secluded a spot, is, I will answer for it, quite dissipated compared with Gower Street. We have given up all parties, for they agree with neither of us; and if one is quiet in London, there is nothing like its quietness— there is a grandeur about its smoky fogs, and the dull distant sounds of cabs and coaches; in fact you may perceive I am becoming a thorough-paced Cockney, and I glory in thoughts that I shall be here for the next six months."

The entries of ill health in the Diary increase in number during these years, and as a consequence the holidays become longer and more frequent. From April 26 to May 13,

* Daughter of Josiah Wedgwood of Maer, and grand-daughter of the founder of the Etruria Works.

1839, he was at Maer and Shrewsbury. Again, from August 23 to October 2 he was away from London at Maer, Shrewsbury, and at Birmingham for the meeting of the British Association.

The entry under August 1839 is: "During my visit to Maer, read a little, was much unwell and scandalously idle. I have derived this much good, that *nothing* is so intolerable as idleness."

At the end of 1839 his eldest child was born, and it was then that he began his observations ultimately published in the 'Expression of the Emotions.' His book on this subject, and the short paper published in 'Mind,'* show how closely he observed his child. He seems to have been surprised at his own feeling for a young baby, for he wrote to Fox (July 1840): "He [*i.e.* the baby] is so charming that I cannot pretend to any modesty. I defy anybody to flatter us on our baby, for I defy any one to say anything in its praise of which we are not fully conscious. . . . I had not the smallest conception there was so much in a five-month baby. You will perceive by this that I have a fine degree of paternal fervour."

During these years he worked intermittently at 'Coral Reefs,' being constantly interrupted by ill health. Thus he speaks of "recommencing" the subject in February 1839, and again in the October of the same year, and once more in July 1841, "after more than thirteen months' interval." His other scientific work consisted of a contribution to the Geological Society,† on the boulders and "till" of South America, as well as a few other minor papers on geological subjects. He also worked busily at the ornithological part of the Zoology of the *Beagle, i.e.* the notice of the habits and ranges of the birds which were described by Gould.

* July 1877.
† 'Geol. Soc. Proc.' iii. 1842, and 'Geol. Soc. Trans.' vi.

C. Darwin to C. Lyell.

Wednesday morning [February 1840].

MY DEAR LYELL,

Many thanks for your kind note. I will send for the *Scotsman.* Dr. Holland thinks he has found out what is the matter with me, and now hopes he shall be able to set me going again. Is it not mortifying, it is now nine weeks since I have done a whole day's work, and not more than four half days. But I won't grumble any more, though it is hard work to prevent doing so. Since receiving your note I have read over my chapter on Coral, and find I am prepared to stand by almost everything ; it is much more cautiously and accurately written than I thought. I had set my heart upon having my volume completed before your new edition, but not, you may believe me, for you to notice anything new in it (for there is very little besides details), but you are the one man in Europe whose opinion of the general truth of a toughish argument I should be always most anxious to hear. My MS. is in such confusion, otherwise I am sure you should most willingly, if it had been worth your while, have looked at any part you choose.

*   *   *   *   *

[In a letter to Fox (January 1841) he shows that his " Species work " was still occupying his mind :—

" If you attend at all to Natural History I send you this P.S. as a memento, that I continue to collect all kinds of facts about ' Varieties and Species,' for my some-day work to be so entitled ; the smallest contributions thankfully accepted ; descriptions of offspring of all crosses between all domestic birds and animals, dogs, cats, &c., &c., very valuable. Don't forget, if your half-bred African cat should die that I should be very much obliged for its carcase sent up in a little hamper for the skeleton ; it, or any cross-bred pigeons, fowl, duck, &c., &c., will be more acceptable than the finest haunch of venison, or the finest turtle."

Later in the year (September) he writes to Fox about his health, and also with reference to his plan of moving into the country :—

" I have steadily been gaining ground, and really believe now I shall some day be quite strong. I write daily for a couple of hours on my Coral volume, and take a little walk or ride every day. I grow very tired in the evenings, and am not able to go out at that time, or hardly to receive my nearest relations ; but my life ceases to be burdensome now that I can do something. We are taking steps to leave London, and live about twenty miles from it on some railway."]

1842.

[The record of work includes his volume on ' Coral Reefs,' * the manuscript of which was at last sent to the printers in January of this year, and the last proof corrected in May. He thus writes of the work in his diary :—

" I commenced this work three years and seven months ago. Out of this period about twenty months (besides work during *Beagle's* voyage) has been spent on it, and besides it, I have only compiled the Bird part of Zoology ; Appendix to Journal, paper on Boulders, and corrected papers on Glen Roy and earthquakes, reading on species, and rest all lost by illness."

In May and June he was at Shrewsbury and Maer, whence he went on to make the little tour in Wales, of which he spoke in his ' Recollections,' and of which the results were published as " Notes on the effects produced by the ancient glaciers of Caernarvonshire, and on the Boulders transported by floating Ice." †

---

* A notice of the Coral Reef work appeared in the ' Geograph. Soc. Journal,' xii. p. 115.

† ' Philosophical Magazine,' 1842, p. 352.

Mr. Archibald Geikie speaks of this paper as standing "almost at the top of the long list of English contributions to the history of the Ice Age." *

The latter part of this year belongs to the period including the settlement at Down, and is therefore dealt with in another chapter.]

* Charles Darwin, 'Nature' Series, p. 23.

CHAPTER VIII.

RELIGION.

[THE history of this part of my father's life may justly include some mention of his religious views. For although, as he points out, he did not give continuous systematic thought to religious questions, yet we know from his own words that about this time (1836–39) the subject was much before his mind.

In his published works he was reticent on the matter of religion, and what he has left on the subject was not written with a view to publication.*

I believe that his reticence arose from several causes. He felt strongly that a man's religion is an essentially private matter, and one concerning himself alone. This is indicated by the following extract from a letter of 1879 :—†

"What my own views may be is a question of no consequence to any one but myself. But, as you ask, I may state that my judgment often fluctuates . . . In my most extreme fluctuations I have never been an Atheist in the sense of denying the existence of a God. I think that generally (and more and more as I grow older), but not always, that an Agnostic would be the more correct description of my state of mind."

* As an exception may be mentioned, a few words of concurrence with Dr. Abbott's ' Truths for the Times,' which my father allowed to be published in the *Index.*

† Addressed to Mr. J. Fordyce, and published by him in his 'Aspects of Scepticism,' 1883.

He naturally shrank from wounding the sensibilities of others in religious matters, and he was also influenced by the consciousness that a man ought not to publish on a subject to which he has not given special and continuous thought. That he felt this caution to apply to himself in the matter of religion is shown in a letter to Dr. F. E. Abbott, of Cambridge, U.S. (Sept. 6, 1871). After explaining that the weakness arising from his bad health prevented him from feeling "equal to deep reflection, on the deepest subject which can fill a man's mind," he goes on to say: "With respect to my former notes to you, I quite forget their contents. I have to write many letters, and can reflect but little on what I write; but I fully believe and hope that I have never written a word, which at the time I' did not think; but I think you will agree with me, that anything which is to be given to the public ought to be maturely weighed and cautiously put. It never occurred to me that you would wish to print any extract from my notes: if it had, I would have kept a copy. I put 'private' from habit, only as yet partially acquired, from some hasty notes of mine having been printed, which were not in the least degree worth printing, though otherwise unobjectionable. It is simply ridiculous to suppose that my former note to you would be worth sending to me, with any part marked which you desire to print; but if you like to do so, I will at once say whether I should have any objection. I feel in some degree unwilling to express myself publicly on religious subjects, as I do not feel that I have thought deeply enough to justify any publicity."

I may also quote from another letter to Dr. Abbott (Nov. 16, 1871), in which my father gives more fully his reasons for not feeling competent to write on religious and moral subjects:—

"I can say with entire truth that I feel honoured by your request that I should become a contributor to the *Index*,

and am much obliged for the draft. I fully, also, subscribe
to the proposition that it is the duty of every one to spread
what he believes to be the truth ; and I honour you for doing
so, with so much devotion and zeal. But I cannot comply
with your request for the following reasons ; and excuse me
for giving them in some detail, as I should be very sorry to
appear in your eyes ungracious. My health is very weak :
I *never* pass 24 hours without many hours of discomfort,
when I can do nothing whatever. I have thus, also, lost two
whole consecutive months this season. Owing to this weakness, and my head being often giddy, I am unable to master
new subjects requiring much thought, and can deal only with
old materials. At no time am I a quick thinker or writer :
whatever I have done in science has solely been by long
pondering, patience and industry.

"Now I have never systematically thought much on religion
in relation to science, or on morals in relation to society ; and
without steadily keeping my mind on such subjects for a
long period, I am really incapable of writing anything worth
sending to the *Index.*"

He was more than once asked to give his views on religion,
and he had, as a rule, no objection to doing so in a private
letter. Thus in answer to a Dutch student, he wrote
(April 2, 1873) :—

"I am sure you will excuse my writing at length, when I
tell you that I have long been much out of health, and am
now staying away from my home for rest.

"It is impossible to answer your question briefly ; and I am
not sure that I could do so, even if I wrote at some length.
But I may say that the impossibility of conceiving that this
grand and wondrous universe, with our conscious selves, arose
through chance, seems to me the chief argument for the
existence of God ; but whether this is an argument of real
value, I have never been able to decide. I am aware that if
we admit a first cause, the mind still craves to know whence

it came, and how it arose. Nor can I overlook the difficulty from the immense amount of suffering through the world. I am, also, induced to defer to a certain extent to the judgment of the many able men who have fully believed in God ;. but here again I see how poor an argument this is. The safest conclusion seems to me that the whole subject is beyond the scope of man's intellect ; but man can do his duty."

Again in 1879 he was applied to by a German student, in a similar manner. The letter was answered by a member of my father's family, who wrote :—

" Mr. Darwin begs me to say that he receives so many letters, that he cannot answer them all.

" He considers that the theory of Evolution is quite compatible with the belief in a God ; but that you must remember that different persons have different definitions of what they mean by God."

This, however, did not satisfy the German youth, who again wrote to my father, and received from him the following reply :—

" I am much engaged, an old man, and out of health, and I cannot spare time to answer your questions fully,—nor indeed can they be answered. Science has nothing to do with Christ, except in so far as the habit of scientific research makes a man cautious in admitting evidence. For myself, I do not believe that there ever has been any revelation. As for a future life, every man must judge for himself between conflicting vague probabilities."

The passages which here follow are extracts, somewhat abbreviated, from a part of the Autobiography, written in 1876, in which my father gives the history of his religious views :—

" During these two years * I was led to think much about religion. Whilst on board the *Beagle* I was quite orthodox,

* Oct. 1836 to Jan. 1839.

and I remember being heartily laughed at by several of the
officers (though themselves orthodox) for quoting the Bible as
an unanswerable authority on some point of morality. I
suppose it was the novelty of the argument that amused
them. But I had gradually come by this time, *i.e.* 1836
to 1839, to see that the Old Testament was no more to be
trusted than the sacred books of the Hindoos. The question
then continually rose before my mind and would not be
banished,—is it credible that if God were now to make a
revelation to the Hindoos, he would permit it to be connected with the belief in Vishnu, Siva, &c., as Christianity is
connected with the Old Testament? This appeared to me
utterly incredible.

" By further reflecting that the clearest evidence would be
requisite to make any sane man believe in the miracles by
which Christianity is supported, — and that the more we
know of the fixed laws of nature the more incredible do
miracles become,—that the men at that time were ignorant
and credulous to a degree almost incomprehensible by us,—
that the Gospels cannot be proved to have been written
simultaneously with the events,—that they differ in many
important details, far too important, as it seemed to me,
to be admitted as the usual inaccuracies of eye-witnesses ;—
by such reflections as these, which I give not as having
the least novelty or value, but as they influenced me, I
gradually came to disbelieve in Christianity as a divine
revelation. The fact that many false religions have spread
over large portions of the earth like wild-fire had some
weight with me.

" But I was very unwilling to give up my belief; I feel
sure of this, for I can well remember often and often inventing
day-dreams of old letters between distinguished Romans, and
manuscripts being discovered at Pompeii or elsewhere, which
confirmed in the most striking manner all that was written in
the Gospels. But I found it more and more difficult, with free

scope given to my imagination, to invent evidence which would suffice to convince me. Thus disbelief crept over me at a very slow rate, but was at last complete. The rate was so slow that I felt no distress.

"Although I did not think much about the existence of a personal God until a considerably later period of my life, I will here give the vague conclusions to which I have been driven. The old argument from design in Nature, as given by Paley, which formerly seemed to me so conclusive, fails, now that the law of natural selection has been discovered. We can no longer argue that, for instance, the beautiful hinge of a bivalve shell must have been made by an intelligent being, like the hinge of a door by man. There seems to be no more design in the variability of organic beings, and in the action of natural selection, than in the course which the wind blows. But I have discussed this subject at the end of my book on the 'Variation of Domesticated Animals and Plants,' * and the argument there given has never, as far as I can see, been answered.

"But passing over the endless beautiful adaptations which we everywhere meet with, it may be asked how can the generally beneficent arrangement of the world be accounted for? Some writers indeed are so much impressed with the amount of suffering in the world, that they doubt, if we look to all sentient beings, whether there is more of misery or of happiness; whether the world as a whole is a good or bad one.

* My father asks whether we are to believe that the forms are preordained of the broken fragments of rock tumbled from a precipice which are fitted together by man to build his houses. If not, why should we believe that the variations of domestic animals or plants are preordained for the sake of the breeder? "But if we give up the principle in one case, . . . no shadow of reason can be assigned for the belief that variations, alike in nature and the result of the same general laws, which have been the groundwork through natural selection of the formation of the most perfectly adapted animals in the world, man included, were intentionally and specially guided."—'The Variation of Animals and Plants,' 1st Edit. vol. ii. p. 431.—F. D.

According to my judgment happiness decidedly prevails, though this would be very difficult to prove. If the truth of this conclusion be granted, it harmonizes well with the effects which we might expect from natural selection. If all the individuals of any species were habitually to suffer to an extreme degree, they would neglect to propagate their kind; but we have no reason to believe that this has ever, or at least often occurred. Some other considerations, moreover, lead to the belief that all sentient beings have been formed so as to enjoy, as a general rule, happiness.

" Every one who believes, as I do, that all the corporeal and mental organs (excepting those which are neither advantageous nor disadvantageous to the possessor) of all beings have been developed through natural selection, or the survival of the fittest, together with use or habit, will admit that these organs have been formed so that their possessors may compete successfully with other beings, and thus increase in number. Now an animal may be led to pursue that course of action which is most beneficial to the species by suffering, such as pain, hunger, thirst, and fear; or by pleasure, as in eating and drinking, and in the propagation of the species, &c.; or by both means combined, as in the search for food. But pain or suffering of any kind, if long continued, causes depression and lessens the power of action, yet is well adapted to make a creature guard itself against any great or sudden evil. Pleasurable sensations, on the other hand, may be long continued without any depressing effect; on the contrary, they stimulate the whole system to increased action. Hence it has come to pass that most or all sentient beings have been developed in such a manner, through natural selection, that pleasurable sensations serve as their habitual guides. We see this in the pleasure from exertion, even occasionally from great exertion of the body or mind,—in the pleasure of our daily meals, and especially in the pleasure derived from sociability, and from loving our families. The sum of such pleasures as these, which are habitual or frequently recurrent, give, as I can hardly doubt, to most sentient beings an excess of happiness over misery, although many occasionally suffer much. Such suffering is quite compatible with the belief in Natural Selection, which is not perfect in its action, but tends only to render each species as successful as possible in the battle for life with other species, in wonderfully complex and changing circumstances.

"That there is much suffering in the world no one disputes. Some have attempted to explain this with reference to man by imagining that it serves for his moral improvement. But the number of men in the world is as nothing compared with that of all other sentient beings, and they often suffer greatly without any moral improvement. This very old argument from the existence of suffering against the existence of an intelligent First Cause seems to me a strong one ; whereas, as just remarked, the presence of much suffering agrees well with the view that all organic beings have been developed through variation and natural selection.

"At the present day the most usual argument for the existence of an intelligent God is drawn from the deep inward conviction and feelings which are experienced by most persons.

"Formerly I was led by feelings such as those just referred to (although I do not think that the religious sentiment was ever strongly developed in me), to the firm conviction of the existence of God, and of the immortality of the soul. In my Journal I wrote that whilst standing in the midst of the grandeur of a Brazilian forest, "it is not possible to give an adequate idea of the higher feelings of wonder, admiration, and devotion, which fill and elevate the mind." I well remember my conviction that there is more in man than the mere breath of his body. But now the grandest scenes would not cause any such convictions and feelings to rise in my mind. It may be truly said that I am like a man who has become colour-blind, and the universal belief by men of the existence of redness makes my present loss of perception of not the least value as evidence. This argument would be a valid one if all men of all races had the same inward conviction of the existence of one God ; but we know that this is very far from being the case. Therefore I cannot see that such inward convictions and feelings are of any weight as evidence of what really exists. The state of mind which grand scenes formerly excited in me, and which was intimately connected with a belief in God, did not essentially differ from that which is often called the sense of sublimity ; and however difficult it may be to explain the genesis of this sense, it can hardly be advanced as an argument for the existence of God, any more than the powerful though vague and similar feelings excited by music.

" With respect to immortality, nothing shows me [so clearly] how strong and almost instinctive a belief it is, as the consideration of the view now held by most physicists, namely, that the sun with all the planets will in time grow too cold for life, unless indeed some great body dashes into the sun and thus gives it fresh life. Believing as I do that man in the distant future will be a far more perfect creature than he now is, it is an intolerable thought that he and all other sentient beings are doomed to complete annihilation after such long-continued slow progress. To those who fully admit the immortality of the human soul, the destruction of our world will not appear so dreadful.

"Another source of conviction in the existence of God, connected with the reason, and not with the feelings, impresses me as having much more weight. This follows from the extreme difficulty or rather impossibility of conceiving this immense and wonderful universe, including man with his capacity of looking far backwards and far into futurity, as the result of blind chance or necessity. When thus reflecting I feel compelled to look to a First Cause having an intelligent mind in some degree analogous to that of man ; and I deserve to be called a Theist. This conclusion was strong in my mind about the time, as far as I can remember, when I wrote the 'Origin of Species ;' and it is since that time that it has very gradually, with many fluctuations, become weaker. But then arises the doubt, can the mind of man, which has, as I fully believe, been developed from a mind as low as that possessed by the lowest animals, be trusted when it draws such grand conclusions ?

" I cannot pretend to throw the least light on such abstruse problems. The mystery of the beginning of all things is insoluble by us ; and I for one must be content to remain an Agnostic."

[The following letters repeat to some extent what has been given from the Autobiography. The first one refers to 'The Boundaries of Science, a Dialogue,' published in 'Macmillan's Magazine,' for July 1861.]

C. Darwin to Miss Julia Wedgwood.

July 11 [1861].

Some one has sent us 'Macmillan'; and I must tell you how much I admire your Article; though at the same time I must confess that I could not clearly follow you in some parts, which probably is in main part due to my not being at all accustomed to metaphysical trains of thought. I think that you understand my book * perfectly, and that I find a very rare event with my critics. The ideas in the last page have several times vaguely crossed my mind. Owing to several correspondents I have been led lately to think, or rather to try to think over some of the chief points discussed by you. But the result has been with me a maze—something like thinking on the origin of evil, to which you allude. The mind refuses to look at this universe, being what it is,

* The 'Origin of Species.'

without having been designed ; yet, where one would most expect design, viz. in the structure of a sentient being, the more I think on the subject, the less I can see proof of design. Asa Gray and some others look at each variation, or at least at each beneficial variation (which A. Gray would compare with the rain drops * which do not fall on the sea, but on to the land to fertilize it) as having been providentially designed. Yet when I ask him whether he looks at each variation in the rock-pigeon, by which man has made by accumulation a pouter or fantail pigeon, as providentially designed for man's amusement, he does not know what to answer ; and if he, or any one, admits [that] these variations are accidental, as far as purpose is concerned (of course not accidental as to their cause or origin) ; then I can see no reason why he should rank the accumulated variations by which the beautifully adapted woodpecker has been formed, as providentially designed. For it would be easy to imagine the enlarged crop of the pouter, or tail of the fantail, as of some use to birds, in a state of nature, having peculiar habits of life. These are the considerations which perplex me about design ; but whether you will care to hear them, I know not.

*　　　*　　　*　　　*　　　*

[On the subject of design, he wrote (July 1860) to Dr. Gray :

"One word more on 'designed laws' and 'undesigned results.' I see a bird which I want for food, take my gun and

* Dr. Gray's rain-drop metaphor occurs in the Essay 'Darwin and his Reviewers' ('Darwiniana,' p. 157): "The whole animate life of a country depends absolutely upon the vegetation, the vegetation upon the rain. The moisture is furnished by the ocean, is raised by the sun's heat from the ocean's surface, and is wafted inland by the winds. But what multitudes of rain-drops fall back into the ocean—are as much without a final cause as the incipient varieties which come to nothing ! Does it therefore follow that the rains which are bestowed upon the soil with such rule and average regularity were not designed to support vegetable and animal life?"

kill it, I do this *designedly*. An innocent and good man stands under a tree and is killed by a flash of lightning. Do you believe (and I really should like to hear) that God *designedly* killed this man? Many or most persons do believe this; I can't and don't. If you believe so, do you believe that when a swallow snaps up a gnat that God designed that that particular swallow should snap up that particular gnat at that particular instant? I believe that the man and the gnat are in the same predicament. If the death of neither man nor gnat are designed, I see no good reason to believe that their *first* birth or production should be necessarily designed."]

C. Darwin to W. Graham.

Down, July 3rd, 1881.

DEAR SIR,

I hope that you will not think it intrusive on my part to thank you heartily for the pleasure which I have derived from reading your admirably written 'Creed of Science,' though I have not yet quite finished it, as now that I am old I read very slowly. It is a very long time since any other book has interested me so much. The work must have cost you several years and much hard labour with full leisure for work. You would not probably expect any one fully to agree with you on so many abstruse subjects; and there are some points in your book which I cannot digest. The chief one is that the existence of so-called natural laws implies purpose. I cannot see this. Not to mention that many expect that the several great laws will some day be found to follow inevitably from some one single law, yet taking the laws as we now know them, and look at the moon, where the law of gravitation—and no doubt of the conservation of energy—of the atomic theory, &c. &c., hold good, and I cannot see that there is then necessarily any purpose. Would there be purpose if the lowest organisms alone, destitute of consciousness existed in the moon? But I have had no practice in abstract reasoning, and I may be all astray. Nevertheless you have expressed my inward conviction, though far more vividly and clearly than I could have done, that the Universe is not the result of chance.* But then with me the horrid doubt always arises whether the convictions of man's mind, which has been developed from the mind of the lower animals, are of any value or at all trustworthy. Would any one trust in the convictions of a monkey's mind, if there are any convictions in such a mind? Secondly, I think that I could make somewhat of a case against the enormous importance which you attribute to our greatest men; I have been accustomed to think, second, third, and fourth rate men of very high importance, at least in the case of Science. Lastly, I could show fight on natural selection having done and doing more for the progress of civilization than you seem inclined to admit. Remember what risk the nations of Europe ran, not so many centuries ago of being overwhelmed by the Turks, and how ridiculous such an idea now is! The more civilized so-called Caucasian races have beaten the Turkish hollow in the struggle for existence. Looking to the world at no very distant date, what an endless number of the lower races will have been eliminated by the higher civilized races throughout the world. But I will write no more, and not even mention the many points in your work which have

---

* The Duke of Argyll (' Good Words,' Ap. 1885, p. 244) has recorded a few words on this subject, spoken by my father in the last year of his life. ". . . in the course of that conversation I said to Mr. Darwin, with reference to some of his own remarkable works on the ' Fertilisation of Orchids,' and upon ' The Earthworms,' and various other observations he made of the wonderful contrivances for certain purposes in nature—I said it was impossible to look at these without seeing that they were the effect and the expression of mind. I shall never forget Mr. Darwin's answer. He looked at me very hard and said, ' Well, that often comes over me with overwhelming force; but at other times," and he shook his head vaguely, adding, " it seems to go away.' "

much interested me. I have indeed cause to apologise for troubling you with my impressions, and my sole excuse is the excitement in my mind which your book has aroused.

I beg leave to remain,
Dear Sir,
Yours faithfully and obliged,
CHARLES DARWIN.

[My father spoke little on these subjects, and I can contribute nothing from my own recollection of his conversation which can add to the impression here given of his attitude towards Religion. Some further idea of his views may, however, be gathered from occasional remarks in his letters.]*

* Dr. Aveling has published an account of a conversation with my father. I think that the readers of this pamphlet ('The Religious Views of Charles Darwin,' Free Thought Publishing Company, 1883) may be misled into seeing more resemblance than really existed between the positions of my father and Dr. Aveling: and I say this in spite of my conviction that Dr. Aveling gives quite fairly his impressions of my father's views. Dr. Aveling tried to show that the terms "Agnostic" and "Atheist" were practically equivalent—that an atheist is one who, without denying the existence of God, is without God, inasmuch as he is unconvinced of the existence of a Deity. My father's replies implied his preference for the unaggressive attitude of an Agnostic. Dr. Aveling seems (p. 5) to regard the absence of aggressiveness in my father's views as distinguishing them in an unessential manner from his own. But, in my judgment, it is precisely differences of this kind which distinguish him so completely from the class of thinkers to which Dr. Aveling belongs.

CHAPTER IX.

LIFE AT DOWN.

1842–1854.

" My life goes on like clockwork, and I am fixed on the spot where I shall end it."

Letter to Captain Fitz-Roy, October, 1846.

[WITH the view of giving, in the next volume, a connected account of the growth of the ' Origin of Species,' I have taken the more important letters bearing on that subject out of their proper chronological position here, and placed them with the rest of the correspondence bearing on the same subject ; so that in the present group of letters we only get occasional hints of the growth of my father's views, and we may suppose ourselves to be looking at his life, as it might have been looked at by those who had no knowledge of the quiet development of his theory of evolution during this period.

On Sept. 14, 1842, my father left London with his family and settled at Down.* In the Autobiographical chapter, his motives for taking this step in the country are briefly given. He speaks of the attendance at scientific societies, and ordinary social duties, as suiting his health so " badly that

---

* I must not omit to mention a member of the household who accompanied him. This was his butler, Joseph Parslow, who remained in the family, a valued friend and servant, for forty years, and became, as Sir Joseph Hooker once remarked to me, " an integral part of the family, and felt to be such by all visitors at the house."

we resolved to live in the country, which we both preferred and have never repented of." His intention of keeping up with scientific life in London is expressed in a letter to Fox (Dec., 1842):—

"I hope by going up to town for a night every fortnight or three weeks, to keep up my communication with scientific men and my own zeal, and so not to turn into a complete Kentish hog."

Visits to London of this kind were kept up for some years at the cost of much exertion on his part. I have often heard him speak of the wearisome drives of ten miles to or from Croydon or Sydenham—the nearest stations—with an old gardener acting as coachman, who drove with great caution and slowness up and down the many hills. In later years, all regular scientific intercourse with London became, as before mentioned, an impossibility.

The choice of Down was rather the result of despair than of actual preference; my father and mother were weary of house-hunting, and the attractive points about the place thus seemed to them to counterbalance its somewhat more obvious faults. It had at least one desideratum, namely quietness. Indeed it would have been difficult to find a more retired place so near to London. In 1842 a coach drive of some twenty miles was the only means of access to Down; and even now that railways have crept closer to it, it is singularly out of the world, with nothing to suggest the neighbour-hood of London, unless it be the dull haze of smoke that sometimes clouds the sky. The village stands in an angle between two of the larger high-roads of the country, one leading to Tunbridge and the other to Westerham and Eden-bridge. It is cut off from the Weald by a line of steep chalk hills on the south, and an abrupt hill, now smoothed down by a cutting and embankment, must formerly have been something of a barrier against encroachments from the side of London. In such a situation, a village, communicating

with the main lines of traffic, only by stony tortuous lanes,
may well have been enabled to preserve its retired character.
Nor is it hard to believe in the smugglers and their strings
of pack-horses making their way up from the lawless old
villages of the Weald, of which the memory still existed
when my father settled in Down. The village stands on
solitary upland country, 500 to 600 feet above the sea,—a
country with little natural beauty, but possessing a certain
charm in the shaws, or straggling strips of wood, capping the
chalky banks and looking down upon the quiet ploughed
lands of the valleys. The village, of three or four hundred
inhabitants, consists of three small streets of cottages meeting
in front of the little flint-built church. It is a place where
new-comers are seldom seen, and the names occurring far
back in the old church registers are still well known in the
village. The smock-frock is not yet quite extinct, though
chiefly used as a ceremonial dress by the "bearers" at
funerals; but as a boy I remember the purple or green
smocks of the men at church.

The house stands a quarter of a mile from the village,
and is built, like so many houses of the last century,
as near as possible to the road—a narrow lane winding
away to the Westerham high-road. In 1842, it was dull
and unattractive enough: a square brick building of three
storeys, covered with shabby whitewash and hanging tiles.
The garden had none of the shrubberies or walls that
now give shelter; it was overlooked from the lane, and was
open, bleak, and desolate. One of my father's first under-
takings was to lower the lane by about two feet, and to build
a flint wall along that part of it which bordered the garden.
The earth thus excavated was used in making banks and
mounds round the lawn : these were planted with evergreens,
which now give to the garden its retired and sheltered
character.

The house was made to look neater by being covered with

THE HOUSE AT DOWN. FROM A DRAWING BY MR. ALFRED PARSONS.
ENGRAVED FOR THE 'CENTURY MAGAZINE,' JANUARY 1883.

To face p. 320, *Vol. I.*

stucco, but the chief improvement effected was the building
of a large bow extending up through three storeys. This bow
became covered with a tangle of creepers, and pleasantly
varied the south side of the house. The drawing-room, with
its verandah opening into the garden, as well as the study in
which my father worked during the later years of his life,
were added at subsequent dates.

Eighteen acres of land were sold with the house, of which
twelve acres on the south side of the house formed a pleasant
field, scattered with fair-sized oaks and ashes. From this
field a strip was cut off and converted into a kitchen garden,
in which the experimental plot of ground was situated, and
where the greenhouses were ultimately put up.

The following letter to Mr. Fox (March 28th, 1843) gives
among other things my father's early impressions of Down :—

" I will tell you all the trifling particulars about myself that
I can think of. We are now exceedingly busy with the first
brick laid down yesterday to an addition to our house ; with
this, with almost making a new kitchen garden and sundry
other projected schemes, my days are very full. I find all
this very bad for geology, but I am very slowly progressing
with a volume, or rather pamphlet, on the volcanic islands
which we visited : I manage only a couple of hours per day,
and that not very regularly. It is uphill work writing books,
which cost money in publishing, and which are not read even
by geologists. I forget whether I ever described this place :
it is a good, very ugly house with 18 acres, situated on a
chalk flat, 560 feet above sea. There are peeps of far distant
country and the scenery is moderately pretty : its chief merit
is its extreme rurality. I think I was never in a more perfectly quiet country. Three miles south of us the great chalk
escarpment quite cuts us off from the low country of Kent,
and between us and the escarpment there is not a village or
gentleman's house, but only great woods and arable fields (the
latter in sadly preponderant numbers), so that we are abso-

lutely at the extreme verge of the world.   The whole country
is intersected by foot-paths ; but the surface over the chalk is
clayey and sticky, which is the worst feature in our purchase.
The dingles and banks often remind me of Cambridgeshire
and walks with you to Cherry Hinton, and other places,
though the general aspect of the country is very different.   I
was looking over my arranged cabinet (the only remnant I
have preserved of all my English insects), and was admiring
Panagæus Crux-major ; it is curious the vivid manner in
which this insect calls up in my mind your appearance, with
little Fan trotting after, when I was first introduced to you.
Those entomological days were very pleasant ones.   I am
very much stronger corporeally, but am little better in being
able to stand mental fatigue, or rather excitement, so that I
cannot dine out or receive visitors, except relations with whom
I can pass some time after dinner in silence."

I could have wished to give here some idea of the position
which, at this period of his life, my father occupied among
scientific men and the reading public generally.   But contemporary notices are few and of no particular value for my
purpose,—which therefore must, in spite of a good deal of
pains, remain unfulfilled.

His 'Journal of Researches' was then the only one of his
books which had any chance of being commonly known.   But
the fact that it was published with the 'Voyages' of Captains
King and Fitz-Roy probably interfered with its general
popularity.   Thus Lyell wrote to him in 1838 ('Lyell's Life,'
ii. p. 43), "I assure you my father is quite enthusiastic about
your journal . . . . and he agrees with me that it would have a
large sale if published separately.   He was disappointed at
hearing that it was to be fettered by the other volumes, for,
although he should equally buy it, he feared so many of
the public would be checked from doing so."   In a notice
of the three voyages in the 'Edinburgh Review' (July,
1839), there is nothing leading a reader to believe that

he would find it more attractive than its fellow-volumes. And, as a fact, it did not become widely known until it was separately published in 1845. It may be noted, however, that the 'Quarterly Review' (December, 1839) called the attention of its readers to the merits of the 'Journal' as a book of travels. The reviewer speaks of the "charm arising from the freshness of heart which is thrown over these virgin pages of a strong intellectual man and an acute and deep observer."

The German translation (1844) of the 'Journal' received a favourable notice in No. 12 of the 'Heidelberger Jahrbücher der Literatur,' 1847—where the Reviewer speaks of the author's "varied canvas, on which he sketches in lively colours the strange customs of those distant regions with their remarkable fauna, flora and geological peculiarities." Alluding to the translation, my father writes—"Dr. Dieffenbach . . . has translated my 'Journal' into German, and I must, with unpardonable vanity, boast that it was at the instigation of Liebig and Humboldt."

The geological work of which he speaks in the above letter to Mr. Fox occupied him for the whole of 1843, and was published in the spring of the following year. It was entitled 'Geological Observations on the Volcanic Islands, visited during the voyage of H.M.S. *Beagle*, together with some brief notices on the geology of Australia and the Cape of Good Hope': it formed the second part of the 'Geology of the Voyage of the *Beagle*,' published "with the Approval of the Lords Commissioners of Her Majesty's Treasury." The volume on 'Coral Reefs' forms Part I. of the series, and was published, as we have seen, in 1842. For the sake of the non-geological reader, I may here quote Professor Geikie's words* on these two volumes—which were up to this time my father's chief geological works. Speaking of the 'Coral Reefs,' he says:—
p. 17, "This well-known treatise, the most original of all its

* Charles Darwin, 'Nature' Series, 1882.

author's geological memoirs, has become one of the classics of geological literature.   The origin of those remarkable rings of coral-rock in mid-ocean has given rise to much speculation, but no satisfactory solution of the problem had been proposed. After visiting many of them, and examining also coral reefs that fringe islands and continents, he offered a theory which for simplicity and grandeur strikes every reader with astonishment.   It is pleasant, after the lapse of many years, to recall the delight with which one first read the 'Coral Reefs'; how one watched the facts being marshalled into their places, nothing being ignored or passed lightly over; and how, step by step, one was led to the grand conclusion of wide oceanic subsidence.   No more admirable example of scientific method was ever given to the world, and even if he had written nothing else, the treatise alone would have placed Darwin in the very front of investigators of nature."

It is interesting to see in the following extract from one of Lyell's letters* how warmly and readily he embraced the theory.   The extract also gives incidentally some idea of the theory itself.

"I am very full of Darwin's new theory of Coral Islands, and have urged Whewell to make him read it at our next meeting.   I must give up my volcanic crater theory for ever, though it cost me a pang at first, for it accounted for so much, the annular form, the central lagoon, the sudden rising of an isolated mountain in a deep sea; all went so well with the notion of submerged, crateriform, and conical volcanoes, . . . and then the fact that in the South Pacific we had scarcely any rocks in the regions of coral islands, save two kinds, coral limestone and volcanic !   Yet spite of all this, the whole theory is knocked on the head, and the annular shape and central lagoon have nothing to do with volcanoes, nor even with a crateriform bottom.   Perhaps Darwin told you when at the

* To Sir John Herschel, May 24, 1837.   'Life of Sir Charles Lyell,' vol. ii. p. 12.

Cape what he considers the true cause ? Let any mountain be submerged gradually, and coral grow in the sea in which it is sinking, and there will be a ring of coral, and finally only a lagoon in the centre. Why? For the same reason that a barrier reef of coral grows along certain coasts : Australia, &c. Coral islands are the last efforts of drowning continents to lift their heads above water. Regions of elevation and subsidence in the ocean may be traced by the state of the coral reefs." There is little to be said as to published contemporary criticism. The book was not reviewed in the ' Quarterly Review ' till 1847, when a favourable notice was given. The reviewer speaks of the "bold and startling " character of the work, but seems to recognize the fact that the views are generally accepted by geologists. By that time the minds of men were becoming more ready to receive geology of this type. Even ten years before, in 1837, Lyell * says, "people are now much better prepared to believe Darwin when he advances proofs of the slow rise of the Andes, than they were in 1830, when I first startled them with that doctrine." This sentence refers to the theory elaborated in my father's geological observations on South America (1846), but the gradual change in receptivity of the geological mind must have been favourable to all his geological work. Nevertheless, Lyell seems at first not to have expected any ready acceptance of the Coral theory ; thus he wrote to my father in 1837 :—" I could think of nothing for days after your lesson on coral reefs, but of the tops of submerged continents. It is all true, but do not flatter yourself that you will be believed till you are growing bald like me, with hard work and vexation at the incredulity of the world."

The second part of the ' Geology of the Voyage of the *Beagle,*' *i.e.* the volume on Volcanic Islands, which specially concerns us now, cannot be better described than by again quoting from Professor Geikie (p. 18) :—

* ' Life of Sir Charles Lyell,' vol. ii. p. 6.

"Full of detailed observations, this work still remains the best authority on the general geological structure of most of the regions it describes. At the time it was written the 'crater of elevation theory,' though opposed by Constant Prévost, Scrope, and Lyell, was generally accepted, at least on the Continent. Darwin, however, could not receive it as a valid explanation of the facts ; and though he did not share the view of its chief opponents, but ventured to propose a hypothesis of his own, the observations impartially made and described by him in this volume must be regarded as having contributed towards the final solution of the difficulty." Professor Geikie continues (p. 21): "He is one of the earliest writers to recognize the magnitude of the denudation to which even recent geological accumulations have been subjected. One of the most impressive lessons to be learnt from his account of 'Volcanic Islands' is the prodigious extent to which they have been denuded. . . . He was disposed to attribute more of this work to the sea than most geologists would now admit; but he lived himself to modify his original views, and on this subject his latest utterances are quite abreast of the time."

An extract from a letter of my father's to Lyell shows his estimate of his own work. "You have pleased me much by saying that you intend looking through my 'Volcanic Islands': it cost me eighteen months !!! and I have heard of very few who have read it. Now I shall feel, whatever little (and little it is) there is confirmatory of old work, or new, will work its effect and not be lost."

The third of his geological books, 'Geological Observations on South America,' may be mentioned here, although it was not published until 1846. "In this work the author embodied all the materials collected by him for the illustration of South American Geology, save some which had been published elsewhere. One of the most important features of the book was the evidence which it brought forward to prove the slow interrupted elevation of the South American Continent during a recent geological period." *

Of this book my father wrote to Lyell :—" My volume will be about 240 pages, dreadfully dull, yet much condensed. I think whenever you have time to look through it, you will think the collection of facts on the elevation of the land and on the formation of terraces pretty good."

Of his special geological work as a whole, Professor Geikie, while pointing out that it was not "of the same epoch-making kind as his biological researches," remarks that he "gave a powerful impulse to " the general reception of Lyell's teaching "by the way in which he gathered from all parts of the world facts in its support."

WORK OF THE PERIOD 1842 TO 1854.

The work of these years may be roughly divided into a period of geology from 1842 to 1846, and one of zoology from 1846 onwards.

I extract from his diary notices of the time spent on his geological books and on his ' Journal.'

'Volcanic Islands.' Summer of 1842 to January, 1844.

'Geology of South America.' July, 1844, to April, 1845.

Second Edition of ' The Journal,' October, 1845, to October, 1846.

The time between October, 1846, and October, 1854, was practically given up to working at the Cirripedia (Barnacles) ; the results were published in two volumes by the Ray Society in 1851 and 1854. His volumes on the Fossil Cirripedes were published by the Palæontographical Society in 1851 and 1854.

Some account of these volumes will be given later.

The minor works may be placed together, independently, of subject matter.

"Observations on the Structure, &c., of the genus Sagitta," Ann. Nat. Hist. xiii., 1844, pp. 1–6.

* Geikie, *loc. cit.*

"Brief Descriptions of several Terrestrial Planariæ, &c.," Ann. Nat. Hist. xiv., 1844, pp. 241–251.

"An Account of the Fine Dust* which often Falls on Vessels in the Atlantic Ocean," Geol. Soc. Journ. ii., 1846, pp. 26–30.

"On the Geology of the Falkland Islands," Geol. Soc. Journ. ii., 1846, pp. 267–274.

"On the Transportal of Erratic Boulders, &c.," Geol. Soc. Journ. iv. 1848, pp. 315–323.†

The article " Geology," in the Admiralty Manual of Scientific Enquiry (1849), pp. 156–195. This was written in the spring of 1848.

"On British Fossil Lepadidæ," 'Geol. Soc. Journ.' vi., 1850, pp. 439–440.

"Analogy of the structure of some Volcanic Rocks with that of Glaciers," 'Edin. Roy. Soc. Proc.' ii., 1851, pp. 17–18.

Professor Geikie has been so good as to give me (in a letter dated Nov. 1885) his impressions of my father's article in the 'Admiralty Manual.' He mentions the following points as characteristic of the work :—

* A sentence occurs in this paper of interest, as showing that the author was alive to the importance of all means of distribution :—" The fact that particles of this size having been brought at least 330 miles from the land is interesting as bearing on the distribution of Cryptogamic plants."

† An extract from a letter to Lyell, 1847, is of interest in connection with this essay :—" Would you be so good (if you know it) as to put Maclaren's address on the enclosed letter and post it. It is chiefly to enquire in what paper he has described the Boulders on Arthur's Seat. Mr. D. Milne in the last Edinburgh ' New Phil. Journal' [1847], has a long paper on it. He says : ' Some glacialists have ventured to explain the transportation of boulders even in the situation of those now referred to, by imagining that they were transported on ice floes,' &c. He treats this view, and the scratching of rocks by icebergs, as almost absurd . . . he has finally stirred me up so, that (without you would answer him) I think I will send a paper in opposition to the same Journal. I can thus introduce some old remarks of mine, and some new, and will insist on your capital observations in N. America. It is a bore to stop one's work, but he has made me quite wroth "

" 1. Great breadth of view. No one who had not practically studied and profoundly reflected on the questions discussed could have written it.

" 2. The insight so remarkable in all that Mr. Darwin ever did. The way in which he points out lines of enquiry that would elucidate geological problems is eminently typical of him. Some of these lines have never yet been adequately followed ; so with regard to them he was in advance of his time.

" 3. Interesting and sympathetic treatment. The author at once puts his readers into harmony with him. He gives them enough of information to show how delightful the field is to which he invites them, and how much they might accomplish in it. There is a broad sketch of the subject which everybody can follow, and there is enough of detail to instruct and guide a beginner and start him on the right track.

" Of course, geology has made great strides since 1849, and the article, if written now, would need to take notice of other branches of enquiry, and to modify statements which are not now quite accurate ; but most of the advice Mr. Darwin gives is as needful and valuable now as when it was given. It is curious to see with what unerring instinct he seems to have fastened on the principles that would stand the test of time."

In a letter to Lyell (1853) my father wrote, " I went up for a paper by the Arctic Dr. Sutherland, on ice action, read only in abstract, but I should think with much good matter. It was very pleasant to hear that it was written owing to the Admiralty Manual."

To give some idea of the retired life which now began for my father at Down, I have noted from his diary the short periods during which he was away from home between the autumn of 1842, when he came to Down, and the end of 1854.

1843, *July*.—Week at Maer and Shrewsbury.

„   *October*.—Twelve days at Shrewsbury.

1844, *April*.—Week at Maer and Shrewsbury.

„   *July*.—Twelve days at Shrewsbury.

1845, *September* 15.—Six weeks, "Shrewsbury, Lincolnshire York, the Dean of Manchester, Waterton, Chatsworth."

1846, *February*.—Eleven days at Shrewsbury.

„   *July*.—Ten days at Shrewsbury.

„   *September*.—Ten days at Southampton, &c., for the British Association.

1847, *February*.—Twelve days at Shrewsbury.

„   *June*.—Ten days at Oxford, &c., for the British Association.

„   *October*.—Fortnight at Shrewsbury.

1848, *May*.—Fortnight at Shrewsbury.

„   *July*.—Week at Swanage.

„   *October*.—Fortnight at Shrewsbury.

„   *November*.—Eleven days at Shrewsbury.

1849, *March to June*.—Sixteen weeks at Malvern.

„   *September*.—Eleven days at Birmingham for the British Association.

1850, *June*.—Week at Malvern.

„   *August*.—Week at Leith Hill, the house of a relative.

„   *October*.—Week at the house of another relative.

1851, *March*.—Week at Malvern.

„   *April*.—Nine days at Malvern.

„   *July*.—Twelve days in London.

1852, *March*.—Week at Rugby and Shrewsbury.

„   *September*.—Six days at the house of a relative.

1853, *July*.—Three weeks at Eastbourne.

„   *August*.—Five days at the military Camp at Chobham.

1854, *March*.—Five days at the house of a relative.

„   *July*.—Three days at the house of a relative.

„   *October*.—Six days at the house of a relative.

It will be seen that he was absent from home sixty weeks in twelve years. But it must be remembered that much of the remaining time spent at Down was lost through ill-health.]

LETTERS.

C. Darwin to R. Fitz-Roy.

Down [March, 31st, 1843].

DEAR FITZ-ROY,—I read yesterday with surprise and the greatest interest, your appointment as Governor of New Zealand. I do not know whether to congratulate you on it, but I am sure I may the Colony, on possessing your zeal and energy. I am most anxious to know whether the report is true, for I cannot bear the thoughts of your leaving the country without seeing you once again ; the past is often in my memory, and I feel that I owe to you much bygone enjoyment, and the whole destiny of my life, which (had my health been stronger) would have been one full of satisfaction to me. During the last three months I have never once gone up to London without intending to call in the hopes of seeing Mrs. Fitz-Roy and yourself ; but I find, most unfortunately for myself, that the little excitement of breaking out of my most quiet routine so generally knocks me up, that I am able to do scarcely anything when in London, and I have not even been able to attend one evening meeting of the Geological Society. Otherwise, I am very well, as are, thank God, my wife and two children. The extreme retirement of this place suits us all very well, and we enjoy our country life much. But I am writing trifles about myself, when your mind and time must be fully occupied. My object in writing is to beg of you or Mrs. Fitz-Roy to have the kindness to send me one line to say whether it is true, and whether you sail soon. I shall come up next week for one or two days ; could you see me for even five minutes, if I called early on Thursday morning, viz. at nine or ten o'clock, or at whatever hour (if you keep early ship hours) you finish your breakfast. Pray remember me very kindly to Mrs. Fitz-Roy, who I trust is able to look at her long voyage with boldness.

Believe me, dear Fitz-Roy,
Your ever truly obliged,
CHARLES DARWIN.

[A quotation from another letter (1846) to Fitz-Roy may be worth giving, as showing my father's affectionate remembrance of his old Captain.

"Farewell, dear Fitz-Roy, I often think of your many acts of kindness to me, and not seldomest on the time, no doubt quite forgotten by you, when, before making Madeira, you came and arranged my hammock with your own hands, and which, as I afterwards heard, brought tears into my father's eyes."]

C. Darwin to W. D. Fox.

[Down, September 5, 1843.]
Monday morning.

MY DEAR FOX,—When I sent off the glacier paper, I was just going out and so had no time to write. I hope your friend will enjoy (and I wish you were going there with him) his tour as much as I did. It was a kind of geological novel. But your friend must have patience, for he will not get a good *glacial eye* for a few days. Murchison and Count Keyserling *rushed* through North Wales the same autumn and could see nothing except the effects of rain trickling over the rocks! I cross-examine Murchison a little, and evidently saw he had looked carefully at nothing. I feel *certain* about the glacier-effects in North Wales. Get up your steam, if this weather lasts, and have a ramble in Wales; its glorious scenery must do every one's heart and body good. I wish I had energy to come to Delamere and go with you; but as you observe, you might as well ask St. Paul's. Whenever I give myself a trip, it shall be, I think, to Scotland, to hunt for more parallel roads. My marine theory for these roads was for a time knocked on the head by Agassiz ice-work, but it is now reviving again. . . . .

Farewell,—we are getting nearly finished—almost all the workmen gone, and the gravel laying down on the walks. Ave Maria! how the money does go. There are twice as many temptations to extravagance in the country compared with London. Adios.

Yours,
C. DARWIN.

C. Darwin to J. D. Hooker.

Down, [1844?]

. . . . I have also read the 'Vestiges,' * but have been somewhat less amused at it than you appear to have been : the writing and arrangement are certainly admirable, but his geology strikes me as bad, and his zoology far worse. I should be very much obliged, if at any future or leisure time you could tell me on what you ground your doubtful belief in imagination of a mother affecting her offspring.† I have attended to the several statements scattered about, but do not

* 'The Vestiges of the Natural History of Creation,' was published anonymously in 1844, and is confidently believed to have been written by the late Robert Chambers. My father's copy gives signs of having been carefully read, a long list of marked passages being pinned in at the end. One useful lesson he seems to have learned from it. He writes : " The idea of a fish passing into a reptile, monstrous. I will not specify any genealogies—much too little known at present." He refers again to the book in a letter to Fox, February, 1845 : " Have you read that strange, unphilosophical, but capitally-written book, the 'Vestiges': it has made more talk than any work of late, and has been by some attributed to me—at which I ought to be much flattered and unflattered."

† This refers to the case of a relative of Sir J. Hooker's, who insisted that a mole, which appeared on one of her children, was the effect of fright upon herself on having, before the birth of the child, blotted with sepia a copy of Turner's 'Liber Studiorum' that had been lent to her with special injunctions to be careful.

believe in more than accidental coincidences. W. Hunter told my father, then in a lying-in hospital, that in many thousand cases, he had asked the mother, *before her confinement,* whether anything had affected her imagination, and recorded the answers; and absolutely not one case came right, though, when the child was anything remarkable, they afterwards made the cap to fit. Reproduction seems governed by such similar laws in the whole animal kingdom, that I am most loth [to believe]. . . .

C. Darwin to J. M. Herbert.

Down, [1844 or 1845].

MY DEAR HERBERT,—I was very glad to see your hand-writing and hear a bit of news about you. Though you cannot come here this autumn, I do hope you and Mrs. Herbert will come in the winter, and we will have lots of talk of old times, and lots of Beethoven.

I have little or rather nothing to say about myself; we live like clock-work, and in what most people would consider the dullest possible manner. I have of late been slaving extra hard, to the great discomfiture of wretched digestive organs, at South America, and thank all the fates, I have done three-fourths of it. Writing plain English grows with me more and more difficult, and never attainable. As for your pre-tending that you will read anything so dull as my pure geological descriptions, lay not such a flattering unction on my soul * for it is incredible. I have long discovered that geologists never read each other's works, and that the only object in writing a book is a proof of earnestness, and that you do not form your opinions without undergoing labour of

---

* On the same subject he wrote to Fitz-Roy: "I have sent my 'South American Geology' to Dover Street, and you will get it, no doubt, in the course of time. You do not know what you threaten when you propose to read it—it is purely geological. I said to my brother, 'You will of course read it,' and his answer was, 'Upon my life, I would sooner even buy it.'"

some kind. Geology is at present very oral, and what I here say is to a great extent quite true. But I am giving you a discussion as long as a chapter in the odious book itself.

I have lately been to Shrewsbury, and found my father surprisingly well and cheerful.

Believe me, my dear old friend, ever yours,

C. DARWIN.

C. Darwin to J. D. Hooker.

Down, Monday [February 10th, 1845].

MY DEAR HOOKER,—I am much obliged for your very agreeable letter ; it was very good-natured, in the midst of your scientific and theatrical dissipation, to think of writing so long a letter to me. I am astonished at your news, and I must condole with you in your *present* view of the Professor-ship,* and most heartily deplore it on my own account. There is something so chilling in a separation of so many hundred miles, though we did not see much of each other when nearer. You will hardly believe how deeply I regret for *myself* your present prospects. I had looked forward to [our] seeing much of each other during our lives. It is a heavy disappointment ; and in a mere selfish point of view, as aiding me in my work, your loss is indeed irreparable. But, on the other hand, I cannot doubt that you take at present a desponding, instead of bright, view of your prospects : surely there are great advantages, as well as disadvantages. The place is one of eminence ; and really it appears to me there are so many indifferent workers, and so few readers, that it is a high advantage, in a purely scientific point of view, for a good worker to hold a position which leads others to attend to his work. I forget whether you attended Edinburgh, as a student, but in my time there was a knot of men who were far from being the indifferent

---

* Sir J. D. Hooker was a candidate for the Professorship of Botany at Edinburgh University.

and dull listeners which you expect for your audience. Reflect what a satisfaction and honour it would be to *make* a good botanist—with your disposition you will be to many what Henslow was at Cambridge to me and others, a most kind friend and guide. Then what a fine garden, and how good a Public Library! why, Forbes always regrets the advantages of Edinburgh for work : think of the inestimable advantage of getting within a short walk of those noble rocks and hills and sandy shores near Edinburgh! Indeed, I cannot pity you much, though I pity myself exceedingly in your loss. Surely lecturing will, in a year or two, with your *great* capacity for work (whatever you may be pleased to say to the contrary) become easy, and you will have a fair time for your Antarctic Flora and general views of distribution. If I thought your Professorship would stop your work, I should wish it and all the good worldly consequences at *el Diavolo*. I know I shall live to see you the first authority in Europe on that grand subject, that almost keystone of the laws of creation, Geographical Distribution. Well, there is one comfort, you will be at Kew, no doubt, every year, so I shall finish by forcing down your throat my sincere congratulations. Thanks for all your news. I grieve to hear Humboldt is failing ; one cannot help feeling, though unrightly, that such an end is humiliating : even when I saw him he talked beyond all reason. If you see him again, pray give him my most respectful and kind compliments, and say that I never forget that my whole course of life is due to having read and re-read as a youth his ' Personal Narrative.' How true and pleasing are all your remarks on his kindness ; think how many opportunities you will have, in your new place, of being a Humboldt to others. Ask him about the river in N.E. Europe, with the Flora very different on its opposite banks. I have got and read your Wilkes ; what a feeble book in matter and style, and how splendidly got up ! Do write me a line from Berlin. Also thanks for the proof-sheets. I did

not, however, mean proof plates ; I value them, as saving me
copying extracts. Farewell, my dear Hooker, with a heavy
heart I wish you joy of your prospects.

Your sincere friend,
C. DARWIN.

[The second edition of the ' Journal,' to which the following
letter refers, was completed between April 25th and August
25th. It was published by Mr. Murray in the ' Colonial and
Home Library,' and in this more accessible form soon had a
large sale.

Up to the time of his first negotiations with Mr. Murray
for its publication in this form, he had received payment only
in the form of a large number of presentation copies, and he
seems to have been glad to sell the copyright of the second
edition to Mr. Murray for 150*l.*

The points of difference between it and the first edition are
of interest chiefly in connection with the growth of the
author's views on evolution, and will be considered later.]

C. Darwin to C. Lyell.

Down [July, 1845].

MY DEAR LYELL,—I send you the first part * of the new
edition [of the ' Journal of Researches '], which I so entirely
owe to you. You will see that I have ventured to dedicate it
to you,† and I trust that this cannot be disagreeable. I have
long wished, not so much for your sake, as for my own feelings
of honesty, to acknowledge more plainly than by mere reference,
how much I geologically owe you. Those authors, however,

---

* No doubt proof-sheets.

† The dedication of the second
edition of the ' Journal of Researches,' is as follows : — " To
Charles Lyell, Esq., F.R.S., this
second edition is dedicated with
grateful pleasure—as an acknowledgment that the chief part of
whatever scientific merit this Journal and the other works of the
Author may possess, has been derived from studying the well-known
and admirable ' Principles of Geology.' "

who like you, educate people's minds as well as teach them special facts, can never, I should think, have full justice done them except by posterity, for the mind thus insensibly improved can hardly perceive its own upward ascent. I had intended putting in the present acknowledgment in the third part of my Geology, but its sale is so exceedingly small that I should not have had the satisfaction of thinking that as far as lay in my power I had owned, though imperfectly, my debt. Pray do not think that I am so silly, as to suppose that my dedication can any ways gratify you, except so far as I trust you will receive it, as a most sincere mark of my gratitude and friendship.   I think I have improved this edition, especially the second part, which I have just finished.   I have added a good deal about the Fuegians, and cut down into half the mercilessly long discussion on climate and glaciers, &c.   I do not recollect anything added to the first part, long enough to call your attention to ; there is a page of description of a very curious breed of oxen in Banda Oriental.   I should like you to read the few last pages ; there is a little discussion on extinction, which will not perhaps strike you as new, though it has so struck me, and has placed in my mind all the difficulties with respect to the causes of extinction, in the same class with other difficulties which are generally quite overlooked and undervalued by naturalists ; I ought, however, to have made my discussion longer and shewn by facts, as I easily could, how steadily every species must be checked in its numbers.

I received your Travels * yesterday ; and I like exceedingly its external and internal appearance ; I read only about a dozen pages last night (for I was tired with hay-making), but I saw quite enough to perceive how *very* much it will interest me, and how many passages will be scored.   I am pleased to find a good sprinkling of Natural History ; I shall be astonished if it does not sell very largely. . . .

* 'Travels in North America,' 2 vols., 1845.

How sorry I am to think that we shall not see you here again for so long ; I wish you may knock yourself a little bit up before you start and require a day's fresh air, before the ocean breezes blow on you. . . .

Ever yours,

C. DARWIN.

C. Darwin to C. Lyell.

Down, Saturday [August 1st, 1845].

MY DEAR LYELL,—I have been wishing to write to you for a week past, but every five minutes' worth of strength has been expended in getting out my second part.* Your note pleased me a good deal more I dare say than my dedication did you, and I thank you much for it. Your work has interested me much, and I will give you my impressions, though, as I never thought you would care to hear what I thought of the non-scientific parts, I made no notes, nor took pains to remember any particular impression of two-thirds of the first volume. The first impression I should say would be with most (though I have literally seen not one soul since reading it) regret at there not being more of the non-scientific [parts]. I am not a good judge, for I have read nothing, *i.e.* non-scientific about North America, but the whole struck me as very new, fresh, and interesting. Your discussions bore to my mind the evident stamp of matured thought, and of conclusions drawn from facts observed by yourself, and not from the opinions of the people whom you met ; and this I suspect is comparatively rare.

Your slave discussion disturbed me much ; but as you would care no more for my opinion on this head than for the ashes of this letter, I will say nothing except that it gave me some sleepless, most uncomfortable hours. Your account of the religious state of the States particularly interested me ; I am surprised throughout at your very proper boldness against

* Of the second edition of the 'Journal of Researches.'

the Clergy. In your University chapter the Clergy, and not
the State of Education, are most severely and justly handled,
and this I think is very bold, for I conceive you might crush
a leaden-headed old Don, as a Don, with more safety, than
touch the finger of that Corporate Animal, the Clergy.   What
a contrast in Education does England shew itself!   Your
apology (using the term, like the old religionists who meant
anything but an apology) for lectures, struck me as very
clever ; but all the arguments in the world on your side, are
not equal to one course of Jamieson's Lectures on the other
side, which I formerly for my sins experienced.   Although I
had read about the ' Coalfields in North America,' I never in
the smallest degree really comprehended their area, their
thickness and favourable position ; nothing hardly astounded
me more in your book.

Some few parts struck me as rather heterogenous, but I do
not know whether to an extent that at all signified.   I missed
however, a good deal, some general heading to the chapters,
such as the two or three principal places visited.   One has no
right to expect an author to write down to the zero of geographical ignorance of the reader ; but I not knowing a single
place, was occasionally rather plagued in tracing your course.
Sometimes in the beginning of a chapter, in one paragraph
your course was traced through a half dozen places ; anyone,
as ignorant as myself, if he could be found, would prefer such
a disturbing paragraph left out.   I cut your map loose, and I
found that a great comfort ; I could not follow your engraved
track.   I think in a second edition, interspaces here and there
of one line open, would be an improvement.   By the way, I
take credit to myself in giving my Journal a less scientific air
in having printed all names of species and genera in Romans ;
the printing looks, also, better.   All the illustrations strike
me as capital, and the map is an admirable volume in itself.
If your 'Principles' had not met with such universal admiration,
I should have feared there would have been too much geology

in this for the general reader; certainly all that the most clear and light style could do, has been done. To myself the geology was an excellent, well-condensed, well-digested *résumé* of all that has been made out in North America, and every geologist ought to be grateful to you. The summing up of the Niagara chapter appeared to me the grandest part ; I was also deeply interested by your discussion on the origin of the Silurian formations. I have made scores of *scores* marking passages hereafter useful to me.

All the coal theory appeared to me very good ; but it is no use going on enumerating in this manner. I wish there had been more Natural History ; I liked *all* the scattered fragments. ' I have now given you an exact transcript of my thoughts, but they are hardly worth your reading. . . .

C. Darwin to C. Lyell.

Down, August 25th [1845].

MY DEAR LYELL,—This is literally the first day on which I have had any time to spare ; and I will amuse myself by beginning a letter to you. . . .

I was delighted with your letter in which you touch on Slavery ; I wish the same feelings had been apparent in your published discussion. But I will not write on this subject, I should perhaps annoy you, and most certainly myself. I have exhaled myself with a paragraph or two in my Journal on the sin of Brazilian slavery ; you perhaps will think that it is in answer to you ; but such is not the case. I have remarked on nothing which I did not hear on the coast of South America. My few sentences, however, are merely an explosion of feeling. How could you relate so placidly that atrocious sentiment * about separating children from their parents ; and in the next page speak of being distressed at the whites not having prospered ; I assure you the contrast

* In the passage referred to, Lyell does not give his own views, but those of a planter.

made me exclaim out. But I have broken my intention, and so no more on this odious deadly subject.

There is a favourable, but not strong enough review on you, in the *Gardeners' Chronicle.* I am sorry to see that Lindley abides by the carbonic acid gas theory. By the way, I was much pleased by Lindley picking out my extinction paragraphs and giving them uncurtailed. To my mind, putting the comparative rarity of existing species in the same category with extinction has removed a great weight ; though of course it does not explain anything, it shows that until we can explain comparative rarity, we ought not to feel any surprise at not explaining extinction. . . .

I am much pleased to hear of the call for a new edition of the 'Principles' : what glorious good that work has done. I fear this time you will not be amongst the old rocks ; how I should rejoice to live to see you publish and discover another stage below the Silurian—it would be the grandest step possible, I think. I am very glad to hear what progress Bunbury is making in fossil Botany ; there is a fine hiatus for him to fill up in this country. I will certainly call on him this winter. · · · From what little I saw of him, I can quite believe everything which you say of his talents. . . .

C. Darwin to J. D. Hooker.

Shrewsbury, [1845 ?]

MY DEAR HOOKER,—I have just received your note, which has astonished me, and has most truly grieved me. I never for one minute doubted of your success, for I most erroneously imagined, that merit was sure to gain the day. I feel most sure that the day will come soon, when those who have voted against you, if they have any shame or conscience in them, will be ashamed at having allowed politics to blind their eyes to your qualifications, and those qualifications vouched for by Humboldt and Brown ! Well, those testimonials must be a consolation to you. *Proh pudor!* I am vexed and indignant by turns. I cannot even take comfort in thinking that I shall see more of you, and extract more knowledge from your well-arranged stock. I am pleased to think, that after having read a few of your letters, I never once doubted the position you will ultimately hold amongst European Botanists. I can think about nothing else, otherwise I should like [to] discuss 'Cosmos'* with you. I trust you will pay me and my wife a visit this autumn at Down. I shall be at Down on the 24th, and till then moving about.

My dear Hooker, allow me to call myself

Your very true friend,

C. DARWIN.

C. Darwin to C. Lyell.

October 8th [1845] Shrewsbury.

. . . I have lately been taking a little tour to see a farm I have purchased in Lincolnshire,† and then to York, where I visited the Dean of Manchester,‡ the great maker of Hybrids, who gave me much curious information. I also visited Waterton at Walton Hall, and was extremely amused at my visit

* A translation of Humboldt's 'Kosmos.'

† He speaks of his Lincolnshire farm in a letter to Henslow (July 4th) :—" I have bought a farm in Lincolnshire, and when I go there this autumn, I mean to see what I can do in providing any cottage on my small estate with gardens. It is a hopeless thing to look to, but I believe few things would do this country more good in future ages than the destruction of primogeniture, so as to lessen the difference in land-wealth, and make more small freeholders. How atrociously unjust are the stamp laws, which render it so expensive for the poor man to buy his quarter of an acre ; it makes one's blood burn with indignation."

‡ Hon. and Rev. W. Herbert. The visit is mentioned in a letter to Dr. Hooker :—" I have been taking a little tour, partly on business, and visited the Dean of Manchester, and had very much interesting talk with him on hybrids, sterility, and variation, &c. &c. He is full of self-gained knowledge, but knows surprisingly little what others have done on the same subjects. He is very heterodox on 'species': not much better, as most naturalists would esteem it, than poor Mr. Vestiges."

there. He is an amusing strange fellow ; at our early dinner,
our party consisted of two Catholic priests and two Mulat-
tresses! He is past sixty years old, and the day before ran
down and caught a leveret in a turnip-field. It is a fine old
house, and the lake swarms with water-fowl. I then saw
Chatsworth, and was in transport with the great hothouse ;
it is a perfect fragment of a tropical forest, and the sight
made me think with delight of old recollections. My little
ten-day tour made me feel wonderfully strong at the time,
but the good effects did not last. My wife, I am sorry to
say, does not get very strong, and the children are the hope
of the family, for they are all happy, life, and spirits. I have
been much interested with Sedgwick's review ; * though I
find it is far from popular with our scientific readers. I think
some few passages savour of the dogmatism of the pulpit, rather
than of the philosophy of the Professor's Chair ; and some of
the wit strikes me as only worthy of ―― in the 'Quarterly.'
Nevertheless, it is a grand piece of argument against muta-
bility of species, and I read it with fear and trembling, but
was well pleased to find that I had not overlooked any of the
arguments, though I had put them to myself as feebly as
milk and water. Have you read ' Cosmos ' yet ? The English
translation is wretched, and the semi-metaphysico-politico
descriptions in the first part are barely intelligible ; but I
think the volcanic discussion well worth your attention, it
has astonished me by its vigour and information. I grieve to
find Humboldt an adorer of Von Buch, with his classification
of volcanos, craters of elevation, &c. &c., and carbonic acid
gas atmosphere. He is indeed a wonderful man.

I hope to get home in a fortnight and stick to my weary-
ful South America till I finish it. I shall be very anxious to
hear how you get on from the Horners, but you must not think
of wasting your time by writing to me. We shall miss, indeed,

* Sedgwick's review of the 'Vestiges of Creation' in the 'Edinburgh
Review,' July 1845.

your visits to Down, and I shall feel a lost man in London without my morning "house of call" at Hart Street. . . .

Believe me, my dear Lyell, ever yours,

C. DARWIN.

C. Darwin to J. D. Hooker.

Down, Farnborough, Kent,
Thursday, September, 1846.

MY DEAR HOOKER,—I hope this letter will catch you at Clifton, but I have been prevented writing by being unwell, and having had the Horners here as visitors, which, with my abominable press-work, has fully occupied my time. It is, indeed, a long time since we wrote to each other; though, I beg to tell you, that I wrote last, but what about I cannot remember, except, I know, it was after reading your last numbers,* and I sent you a uniquely laudatory epistle, considering it was from a man who hardly knows a Daisy from a Dandelion to a professed Botanist. . . .

I cannot remember what papers have given me the impression, but I have that, which you state to be the case, firmly fixed on my mind, namely, the little chemical importance of the soil to its vegetation. What a strong fact it is, as R. Brown once remarked to me, of certain plants being calcareous ones here, which are not so under a more favourable climate on the Continent, or the reverse, for I forget which; but you, no doubt, will know to what I refer. By-the-way, there are some such cases in Herbert's paper in the 'Horticultural Journal.'† Have you read it: it struck me as extremely original, and bears *directly* on your present researches.‡ To a *non-botanist* the chalk has the most peculiar aspect of any flora in England; why will you not come here to make your observations? *We* go to Southampton, if my

---

* Hooker's Antarctic Botany.
† 'Journal of the Horticultural Society,' 1846.

‡ Sir J. Hooker was at this time attending to polymorphism, variability, &c.

courage and stomach do not fail, for the Brit. Assoc. (Do
you not consider it your duty to be there?) And why cannot
you come here afterwards and *work ?* . . . .

THE MONOGRAPH OF THE CIRRIPEDIA,

October 1846 *to October* 1854.

[Writing to Sir J. D. Hooker in 1845, my father says : " I
hope this next summer to finish my South American Geology,
then to get out a little Zoology, and hurrah for my species
work. . ." This passage serves to show that he had at this
time no intention of making an exhaustive study of the
Cirripedes. Indeed it would seem that his original intention
was, as I learn from Sir J. D. Hooker, merely to work out one
special problem. This is quite in keeping with the following
passage in the Autobiography : " When on the coast of Chile,
I found a most curious form, which burrowed into the shells
of Concholepas, and which differed so much from all other
Cirripedes that I had to form a new sub-order for its sole
reception. . . . To understand the structure of my new
Cirripede I had to examine and dissect many of the common forms ; and this gradually led me on to take up the
whole group." In later years he seems to have felt some
doubt as to the value of these eight years of work,—for
instance when he wrote in his Autobiography—" My work
was of considerable use to me, when I had to discuss in the
' Origin of Species' the principles of a natural classification.
Nevertheless I doubt whether the work was worth the consumption of so much time." Yet I learn from Sir J. D.
Hooker that he certainly recognised at the time its value to
himself as systematic training. Sir Joseph writes to me :
" Your father recognised three stages in his career as a
biologist : the mere collector at Cambridge ; the collector and
observer in the *Beagle,* and for some years afterwards ; and
the trained naturalist after, and only after the Cirripede

work. That he was a thinker all along is true enough, and there is a vast deal in his writings previous to the Cirripedes that a trained naturalist could but emulate. . . . He often alluded to it as a valued discipline, and added that even the 'hateful' work of digging out synonyms, and of describing, not only improved his methods but opened his eyes to the difficulties and merits of the works of the dullest of cataloguers. One result was that he would never allow a depreciatory remark to pass unchallenged on the poorest class of scientific workers, provided that their work was honest, and good of its kind. I have always regarded it as one of the finest traits of his character,—this generous appreciation of the hod-men of science, and of their labours . . . and it was monographing the Barnacles that brought it about."

Professor Huxley allows me to quote his opinion as to the value of the eight years given to the Cirripedes :—

"In my opinion your sagacious father never did a wiser thing than when he devoted himself to the years of patient toil which the Cirripede-book cost him.

"Like the rest of us, he had no proper training in biological science, and it has always struck me as a remarkable instance of his scientific insight, that he saw the necessity of giving himself such training, and of his courage, that he did not shirk the labour of obtaining it.

"The great danger which besets all men of large speculative faculty, is the temptation to deal with the accepted statements of fact in natural science, as if they were not only correct, but exhaustive; as if they might be dealt with deductively, in the same way as propositions in Euclid may be dealt with. In reality, every such statement, however true it may be, is true only relatively to the means of observation and the point of view of those who have enunciated it. So far it may be depended upon. But whether it will bear every speculative conclusion that may be logically deduced from it, is quite another question.

" Your father was building a vast superstructure upon the
foundations furnished by the recognised facts of geological
and biological science. In Physical Geography, in Geology
proper, in Geographical Distribution, and in Palæontology, he
had acquired an extensive practical training during the
voyage of the *Beagle*. He knew of his own knowledge the
way in which the raw materials of these branches of science
are acquired, and was therefore a most competent judge of
the speculative strain they would bear. That which he
needed, after his return to England, was a corresponding
acquaintance with Anatomy and Development, and their relation to Taxonomy—and he acquired this by his Cirripede
work.

" Thus, in my apprehension, the value of the Cirripede
monograph lies not merely in the fact that it is a very admirable piece of work, and constituted a great addition to
positive knowledge, but still more in the circumstance that it
was a piece of critical self-discipline, the effect of which mani-
fested itself in everything your father wrote afterwards, and
saved him from endless errors of detail.

" So far from such work being a loss of time, I believe it
would have been well worth his while, had it been prac-
ticable, to have supplemented it by a special study of embryology and physiology. His hands would have been
greatly strengthened thereby when he came to write out
sundry chapters of the ' Origin of Species.' But of course in
those days it was almost impossible for him to find facilities
for such work."

No one can look at the two volumes on the recent Cirripedes, of 399 and 684 pages respectively (not to speak of
the volumes on the fossil species), without being struck
by the immense amount of detailed work which they contain. The forty plates, some of them with thirty figures,
and the fourteen pages of index in the two volumes together, give some rough idea of the labour spent on the

work.* The state of knowledge, as regards the Cirripedes, was most unsatisfactory at the time that my father began to work at them. As an illustration of this fact, it may be mentioned that he had even to re-organise the nomenclature of the group, or, as he expressed it, he " unwillingly found it indispensable to give names to several valves, and to some few of the softer parts of Cirripedes." † It is interesting to learn from his diary the amount of time which he gave to different genera. Thus the genus Chthamalus, the description of which occupies twenty-two pages, occupied him for thirty-six days ; Coronula took nineteen days, and is described in twenty-seven pages. Writing to Fitz-Roy, he speaks of being " for the last half-month daily hard at work in dissecting a little animal about the size of a pin's head, from the Chonos archipelago, and I could spend another month, and daily see more beautiful structure."

Though he became excessively weary of the work before the end of the eight years, he had much keen enjoyment in the course of it. Thus he wrote to Sir J. D. Hooker (1847 ?) : —"As you say, there is an extraordinary pleasure in pure observation ; not but what I suspect the pleasure in this case is rather derived from comparisons forming in one's mind with allied structures. After having been so long employed in writing my old geological observations, it is delightful to use one's eyes and fingers again." It was, in fact, a return to the work which occupied so much of his time when at sea during his voyage. His zoological notes of that period give an impression of vigorous work, hampered by ignorance and want of appliances ; and his untiring industry in the dissection of marine animals, especially of Crustacea, must have been of value to him as training for his Cirripede work. Most of his work was done with the simple dissecting micro-

* The reader unacquainted with Zoology will find some account of the more interesting results in Mr.   Romanes' article on " Charles Darwin " ('Nature' Series, 1882).   † Vol. i. p. 3.

scope—but it was the need which he found for higher powers
that induced him, in 1846, to buy a compound microscope.
He wrote to Hooker:—"When I was drawing with L., I
was so delighted with the appearance of the objects, especially
with their perspective, as seen through the weak powers of a
good compound microscope, that I am going to order one ;
indeed, I often have structures in which the $\frac{1}{30}$ is not power
enough."

During part of the time covered by the present chapter, my
father suffered perhaps more from ill-health than at any other
time of his life. He felt severely the depressing influence of
these long years of illness ; thus as early as 1840 he wrote to
Fox : " I am grown a dull, old, spiritless dog to what I used
to be. One gets stupider as one grows older I think." It is
not wonderful that he should so have written, it is rather
to be wondered at that his spirit withstood so great and
constant a strain. He wrote to Sir J. Hooker in 1845 :
" You are very kind in your enquiries about my health ; I
have nothing to say about it, being always much the same,
some days better and some worse. I believe I have not
had one whole day, or rather night, without my stomach
having been greatly disordered, during the last three years,
and most days great prostration of strength : thank you for
your kindness ; many of my friends, I believe, think me a
hypochondriac."

Again, in 1849, he notes in his diary :—" January 1st to
March 10th. — Health very bad, with much sickness and
failure of power. Worked on all well days." This was
written just before his first visit to Dr. Gully's Water-Cure
Establishment at Malvern. In April of the same year he
wrote :—" I believe I am going on very well, but I am rather
weary of my present inactive life, and the water-cure has the
most extraordinary effect in producing indolence and stagna-
tion of mind : till experiencing it, I could not have believed
it possible. I now increase in weight, have escaped sickness

for thirty days." He returned in June, after sixteen weeks' absence, much improved in health, and, as already described (p. 131), continued the water-cure at home for some time.]

C. Darwin to J. D. Hooker.

Down [October, 1846].

MY DEAR HOOKER,— I have not heard from Sulivan * lately; when he last wrote he named from 8th to 10th as the most likely time. Immediately that I hear, I will fly you a line, for the chance of your being able to come. I forget whether you know him, but I suppose so; he is a real good fellow. Anyhow, if you do not come then, I am very glad that you propose coming soon after. . . .

I am going to begin some papers on the lower marine animals, which will last me some months, perhaps a year, and then I shall begin looking over my ten-year-long accumulation of notes on species and varieties, which, with writing, I dare say will take me five years, and then, when published, I dare say I shall stand infinitely low in the opinion of all sound Naturalists—so this is my prospect for the future.

Are you a good hand at inventing names. I have a quite new and curious genus of Barnacle, which I want to name, and how to invent a name completely puzzles me.

By the way, I have told you nothing about Southampton. We enjoyed (wife and myself) our week beyond measure : the papers were all dull, but I met so many friends and made so many new acquaintances (especially some of the Irish Naturalists), and took so many pleasant excursions. I wish you had been there. On Sunday we had so pleasant an excursion to Winchester with Falconer,† Colonel

* Admiral Sir B. J. Sulivan, formerly an officer of the *Beagle*.

† Hugh Falconer, born 1809, died 1865. Chiefly known as a palæontologist, although employed as a botanist during his whole career in India, where he was also a medical officer in H.E.I.C. Service ; he was superintendent of the Company's garden, first at Saha-

Sabine,* and Dr. Robinson,† and others. I never enjoyed
a day more in my life. I missed having a look at H.
Watson.‡ I suppose you heard that he met Forbes and told
him he had a severe article in the Press. I understand that
Forbes explained to him that he had no cause to complain,
but as the article was printed, he would not withdraw it, but
offered it to Forbes for him to append notes to it, which
Forbes naturally declined. . . .

C. Darwin to J. D. Hooker.

Down, April 7th, [1847 ?]

MY DEAR HOOKER,—I should have written before now, had
I not been almost continually unwell, and at present I am
suffering from four boils and swellings, one of which hardly
allows me the use of my right arm, and has stopped all my
work, and damped all my spirits. I was much disappointed
at missing my trip to Kew, and the more so, as I had forgotten
you would be away all this month; but I had no choice, and
was in bed nearly all Friday and Saturday. I congratulate
you over your improved prospects about India,§ but at the

---

runpore, and then at Calcutta. He
was one of the first botanical ex-
plorers of Kashmir. Falconer's
discoveries of Miocene mammalian
remains in the Sewalik Hills, were,
at the time, perhaps the greatest
"finds" which had been made. His
book on the subject, 'Fauna An-
tiqua Sivalensis,' remained un-
finished at the time of his death.

* The late Sir Edward Sabine,
formerly President of the Royal
Society, and author of a long
series of memoirs on Terrestrial
Magnetism.

† The late Dr. Thomas Romney

Robinson, of the Armagh Observatory.

‡ The late Hewett Cottrell Watson, author of the 'Cybele Britan-
nica,' one of a most valuable series
of works on the topography and
geographical distribution of the
plants of the British Islands.

§ Sir J. Hooker left England on
November 11, 1847, for his Himalayan and Tibetan journey. The
expedition was supported by a small
grant from the Treasury, and thus
assumed the character of a Government mission.

same time must sincerely groan over it.   I shall feel quite lost
without you to discuss many points with, and to point out
(ill-luck to you) difficulties and objections to my species hypotheses.   It will be a horrid shame if money stops your expedition ; but Government will surely help you to some extent.
. . . Your present trip, with your new views, amongst the
coal-plants, will be very interesting.   If you have spare time,
but not without, I should enjoy having some news of your
progress.   Your present trip will work well in, if you go to
any of the coal districts in India.   Would this not be a good
object to parade before Government ; their utilitarian souls
would comprehend this.   By the way, I will get some work
out of you, about the domestic races of animals in India. . . .

C. Darwin to L. Jenyns (Blomefield).

Down [1847].

DEAR JENYNS,—I am very much obliged for the capital
little Almanack ; * it so happened that I was wishing for one
to keep in my portfolio.   I had never seen this kind before,
and shall certainly get one for the future.   I think it is very
amusing to have a list before one's eyes of the order of appearance of the plants and animals around one ; it gives a fresh
interest to each fine day.   There is one point I should like to
see a little improved, viz. the correction for the clock at

* " This letter relates to a small
Almanack first published in 1843,
under the name of ' The Naturalists'
Pocket Almanack,' by Mr. Van
Voorst, and which I edited for him.
It was intended especially for those
who interest themselves in the
periodic phenomena of animals
and plants, of which a select list
was given under each month of
the year.
" The Pocket Almanack contained, moreover, miscellaneous information relating to Zoology and
Botany ; to Natural History and
other scientific societies ; to public
Museums and Gardens, in addition
to the ordinary celestial phenomena
found in most other Almanacks.
It continued to be issued till 1847,
after which year the publication
was abandoned."—From a letter
from Rev. L. Blomefield to F.
Darwin.

shorter intervals. Most people, I suspect, who like myself have dials, will wish to be more precise than with a margin of three minutes. I always buy a shilling almanack for this *sole* end. By the way, *yours, i.e.* Van Voorst's Almanack, is very dear ; it ought, at least, to be advertised post-free for the shilling. Do you not think a table (not rules) of conversion of French into English measures, and perhaps weights, would be exceedingly useful ; also centigrade into Fahrenheit,—magnifying powers according to focal distances?—in fact you might make it the most useful publication of the age. I know what I should like best of all, namely, current meteorological remarks for each month, with statement of average course of winds and prediction of weather, in accordance with movements of barometer. People, I think, are always amused at knowing the extremes and means of temperature for corresponding times in other years.

I hope you will go on with it another year. With many thanks, my dear Jenyns,

Yours very truly,
CHARLES DARWIN.

C. Darwin to J. D. Hooker.

Down, Sunday [April 18th, 1847].

MY DEAR HOOKER,—I return with many thanks Watson's letter, which I have had copied. It is a capital one, and I am extremely obliged to you for obtaining me such valuable information. Surely he is rather in a hurry when he says intermediate varieties must almost be necessarily rare, otherwise they would be taken as the types of the species ; for he overlooks numerical frequency as an element. Surely if A, B, C were three varieties, and if A were a good deal the commonest (therefore, also, first known), it would be taken as the type, without regarding whether B was quite intermediate or not, or whether it was rare or not. What capital essays W. would write ; but I suppose he has written a good deal in the

'Phytologist.' You ought to encourage him to publish on variation ; it is a shame that such facts as those in his letter should remain unpublished. I must get you to introduce me to him ; would he be a good and sociable man for Dropmore ?* though if he comes, Forbes must not (and I think you talked of inviting Forbes), or we shall have a glorious battle. I should like to see sometime the war correspondence. Have you the 'Phytologist,' and could you sometime spare it ; I would go through it quickly. . . . I have read your last five numbers,† and as usual have been much interested in several points, especially with your discussions on the beech and potato. I see you have introduced several sentences against us Transmutationists. I have also been looking through the latter volumes of the 'Annals of Natural History,' and have read two such soulless, pompous papers of ——, quite worthy of the author. . . . The contrast of the papers in the *Annals* with those in the *Annales* is rather humiliating ; so many papers in the former, with short descriptions of species, without one word on their affinities, internal structure, range, or habits. I am now reading ——, and I have picked out some things which have interested me ; but he strikes me as rather dullish, and with all his Materia Medica smells of the doctor's shop. I shall ever hate the name of the Materia Medica, since hearing Duncan's lectures at eight o'clock on a winter's morning—a whole, cold, breakfastless hour on the properties of rhubarb !

I hope your journey will be very prosperous. Believe me, my dear Hooker,

Ever yours,
C. DARWIN.

P.S.—I think I have only made one new acquaintance of late, that is, R. Chambers ; and I have just received a

---

presentation copy of the sixth edition of the 'Vestiges.' Some-how I now feel perfectly convinced he is the author. He is in France, and has written to me thence.

C. Darwin to J. D. Hooker.

Down, [1847 ?]

. . . I am delighted to hear that Brongniart thought Sigillaria aquatic, and that Binney considers coal a sort of submarine peat. I would bet 5 to 1 that in twenty years this will be generally admitted;* and I do not care for whatever the botanical difficulties or impossibilities may be. If I could but persuade myself that Sigillaria and Co. had a good range of depth, *i.e.* could live from 5 to 100 fathoms under water, all difficulties of nearly all kinds would be re-moved (for the simple fact of muddy ordinary shallow sea mplies proximity of land). [N.B.—I am chuckling to think how you are sneering all this time.] It is not much of a difficulty, there not being shells with the coal, considering how unfavourable deep mud is for most Mollusca, and that shells would probably decay from the humic acid, as seems to take place in peat and in the *black* moulds (as Lyell tells me) of the Mississippi. So coal question settled—Q. E. D. Sneer away!

Many thanks for your welcome note from Cambridge, and I am glad you like my *alma mater*, which I despise heartily as a place of education, but love from many most pleasant recollections. . . .

Thanks for your offer of the ' Phytologist ;' I shall be very much obliged for it, for I do not suppose I should be able to borrow it from any other quarter. I will not be set up too much by your praise, but I do not believe I ever lost a book or forgot to return it during a long lapse of time. Your ' Webb' is well wrapped up, and with your name in large letters *outside*.

* An unfulfilled prophecy.

My new microscope is come home (a "splendid plaything," as old R. Brown called it), and I am delighted with it ; it really is a splendid plaything. I have been in London for three days, and saw many of our friends. I was extremely sorry to hear a not very good account of Sir William. Farewell, my dear Hooker, and be a good boy, and make Sigillaria a submarine sea-weed.

Ever yours,

C. DARWIN.

C. Darwin to J. D. Hooker.

Down [May 6th, 1847].

MY DEAR HOOKER,—You have made a savage onslaught, and I must try to defend myself. But, first, let me say that I never write to you except for my own good pleasure ; now I fear that you answer me when busy and without inclination (and I am·sure I should have none if I was as busy as you). Pray do not do so, and if I thought my writing entailed an answer from you *nolens volens,* it would destroy all my pleasure in writing. Firstly, I did not consider my letter as *reasoning,* or even as *speculation,* but simply as mental rioting ; and as I was sending Binney's paper, I poured out to you the result of reading it. Secondly, you are right, indeed, in thinking me mad, if you suppose that I would class any ferns as marine plants ; but surely there is a wide distinction between the plants found upright in the coal-beds and those not upright, and which might have been drifted. Is it not possible that the same circumstances which have preserved the vegetation *in situ,* should have preserved drifted plants ? I know Calamites is found upright ; but I fancied its affinities were very obscure, like Sigillaria. As for Lepidodendron, I forgot its existence, as happens when one goes riot, and now know neither what it is, or whether upright. If these plants, *i.e.* Calamites and Lepidodendron, have *very clear relations* to terrestrial vegetables, like the ferns have, and are found

2 B 2

upright *in situ*, of course I must give up the ghost. But surely Sigillaria is the main upright plant, and on its obscure affinities I have heard you enlarge.

Thirdly, it never entered my head to undervalue botanical relatively to zoological evidence ; except in so far as I thought it was admitted that the vegetative structure seldom yielded any evidence of affinity nearer than that of families, and not always so much. And is it not in plants, as certainly it is in animals, dangerous to judge of habits without very near affinity. Could a Botanist tell from structure alone that the Mangrove family, almost or quite alone in Dicotyledons, could live in the sea, and the Zostera family almost alone among the Monocotyledons? Is it a safe argument, that because algæ are almost the only, or the only submerged sea-plants, that formerly other groups had not members with such habits? With animals such an argument would not be conclusive, as I could illustrate by many examples ; but I am forgetting myself ; I want only to some degree to defend myself, and not burn my fingers by attacking you. The foundation of my letter, and what is my deliberate opinion, though I dare say you will think it absurd, is that I would rather trust, *cæteris paribus*, pure geological evidence than either zoological or botanical evidence. I do not say that I would sooner trust *poor* geological evidence than *good* organic. I think the basis of pure geological reasoning is simpler (consisting chiefly of the action of water on the crust of the earth, and its up and down movements) than a basis drawn from the difficult subject of affinities and of structure in relation to habits. I can hardly analyse the facts on which I have come to this conclusion ; but I can illustrate it. Pallas's account would lead any one to suppose that the Siberian strata, with the frozen carcasses, had been quickly deposited, and hence that the embedded animals had lived in the neighbourhood ; but our zoological knowledge of thirty years ago led every one falsely to reject this conclusion.

Tell me that an upright fern *in situ* occurs with Sigillaria and Stigmaria, or that the affinities of Calamites and Lepidodendron (supposing that they are found *in situ* with Sigillaria) are so *clear*, that they could not have been marine, like, but in a greater degree, than the mangrove and seawrack, and I will humbly apologise to you and all Botanists for having let my mind run riot on a subject on which assuredly I know nothing. But till I hear this, I shall keep privately to my own opinion with the same pertinacity and, as you will think, with the same philosophical spirit with which Koenig maintains that Cheirotherium-footsteps are fuci.

Whether this letter will sink me still lower in your opinion, or put me a little right, I know not, but hope the latter. Anyhow, I have revenged myself with boring you with a very long epistle. Farewell, and be forgiving. Ever yours,

C. DARWIN.

P.S.—When will you return to Kew? I have forgotten one main object of my letter, to thank you *much* for your offer of the 'Hort. Journal,' but I have ordered the two numbers.

[The two following extracts [1847] give the continuation and conclusion of the coal battle.

"By the way, as submarine coal made you so wrath, I thought I would experimentise on Falconer and Bunbury* together, and it made [them] even more savage ; 'such infernal nonsense ought to be thrashed out of me.' Bunbury was more polite and contemptuous. So I now know how to stir up and show off any Botanist. I wonder whether Zoologists and Geologists have got their tender points ; I wish I could find out."

"I cannot resist thanking you for your most kind note. Pray do not think that I was annoyed by your letter : I perceived that you had been thinking with animation, and accordingly expressed yourself strongly, and so I understood it.

* The late Sir C. Bunbury, well known as a palæobotanist.

Forfend me from a man who weighs every expression with Scotch prudence. I heartily wish you all success in your noble problem, and I shall be very curious to have some talk with you and hear your ultimatum."]

C. Darwin to J. D. Hooker.*

Down [October, 1847].

I congratulate you heartily on your arrangements being completed, with some prospect for the future. It will be a noble voyage ·and journey, but I wish it was over, I shall miss you selfishly and all ways to a dreadful extent ... I am in great perplexity how we are to meet ... I can well understand how dreadfully busy you must be. If you *cannot* come here, you *must* let me come to you for a night ; for I must have one more chat and one more quarrel with you over the coal.

By the way, I endeavoured to stir up Lyell (who has been staying here some days with me) to theorise on the coal : his oolitic *upright* Equisetums are dreadful for my submarine flora. I should die much easier if some one would solve me the coal question. I sometimes think it could not have been formed at all. Old Sir Anthony Carlisle once said to me gravely, that he supposed Megatherium and such cattle were just sent down from heaven to see whether the earth would support them ; and I suppose the coal was rained down to puzzle mortals. You must work the coal well in India.

Ever yours,

C. DARWIN.

C. Darwin to J. D. Hooker.

[November 6th, 1847.]

MY DEAR HOOKER,—I have just received your note with sincere grief : there is no help for it. I shall always look at your intention of coming here, under such circumstances, as

* Parts of two letters.

the greatest proof of friendship I ever received from mortal man. My conscience would have upbraided me in not having come to you on Thursday, but, as it turned out, I could not, for I was quite unable to leave Shrewsbury before that day, and I reached home only last night, much knocked up. Without I hear to-morrow (which is hardly possible), and if I am feeling pretty well, I will drive over to Kew on Monday morning, just to say farewell. I will stay only an hour. . . .

C. Darwin to J. D. Hooker.

[November 1847.]

MY DEAR HOOKER,—I am very unwell, and incapable of doing anything. I do hope I have not inconvenienced you. I was so unwell all yesterday, that I was rejoicing you were not here ; for it would have been a bitter mortification to me to have had you here and not enjoyed your last day. I shall not now see you. Farewell, and God bless you.

Your affectionate friend,

C. DARWIN.

I will write to you in India.

[In 1847 appeared a paper by Mr. D. Milne,* in which my father's Glen Roy work is criticised, and which is referred to in the following characteristic extract from a letter to Sir J. Hooker : " I have been bad enough for these few last days, having had to think and write too much about Glen Roy. . . . Mr. Milne having attacked my theory, which made me horribly sick." I have not been able to find any published reply to Mr. Milne, so that I imagine the " writing " mentioned was confined to letters. Mr. Milne's paper was not destructive to the Glen Roy paper, and this my father recognises in the following extract from a letter to Lyell (March, 1847). The reference to Chambers is explained by the fact that he ac-

* Now Mr. Milne Home. The essay was published in Transactions of the Edinburgh Royal Society, vol. xvi.

companied Mr. Milne in his visit to Glen Roy. " I got R.
Chambers to give me a sketch of Milne's Glen Roy views,
and I have re-read my paper, and am, now that I have heard
what is to be said, not even staggered. It is provoking and
humiliating to find that Chambers not only had not read
with any care my paper on this subject, or even looked at the
coloured map, so that the new shelf described by me had not
been searched for, and my arguments and facts of detail not in
the least attended to. I entirely gave up the ghost, and was
quite chicken-hearted at the Geological Society, till you
reassured and reminded me of the main facts in the whole
case."

The two following letters to Lyell, though of later date
(June, 1848), bear on the same subject :—

"I was at the evening meeting [of the Geological Society],
but did not get within hail of you. What a fool (though I must
say a very amusing one) —— did make of himself. Your
speech was refreshing after it, and was well characterized by
Fox (my cousin) in three words—'What a contrast!' That
struck me as a capital speculation about the Wealden Continent going down. I did not hear what you settled at the
Council ; I was quite wearied out and bewildered. I find Smith,
of Jordan Hill, has a much worse opinion of R. Chambers's
book than even I have. Chambers has piqued me a little ; *
he says I ' propound ' and ' profess my belief ' that Glen Roy is
marine, and that the idea was accepted because the ' mobility
of the land was the ascendant idea of the day.' He adds some
very faint *upper* lines in Glen Spean (seen, by the way, by
Agassiz), and has shown that Milne and Kemp are right in
there being horizontal aqueous markings (*not* at coincident
levels with those of Glen Roy) in other parts of Scotland at
great heights, and he adds several other cases. This is the
whole of his addition to the data. He not only takes my line

---

* 'Ancient Sea Margins, 1848.'   should be " the mobility of the land
The words quoted by my father   was an ascendant idea."

of argument from the buttresses and terraces below the lower shelf and some other arguments (without acknowledgment), but he sneers at all his predecessors not having perceived the importance of the short portions of lines intermediate between the chief ones in Glen Roy ; whereas I commence the description of them with saying, that ' perceiving their importance, I examined them with scrupulous care,' and expatiate at considerable length on them.   I have indirectly told him I do not think he has quite claims to consider that he alone (which he pretty directly asserts) has solved the problem of Glen Roy. With respect to the terraces at lower levels coincident in height all round Scotland and England, I am inclined to believe he shows some little probability of there being some leading ones coincident, but much more exact evidence is required.   Would you believe it credible ? he advances as a probable solution to account for the rise of Great Britain that in some great ocean one-twentieth of the bottom of the whole aqueous surface of the globe has sunk in (he does not say where he puts it) for a thickness of half a mile, and this he has calculated would make an apparent rise of 130 feet."

C. Darwin to C. Lyell.

Down [June 1848].

MY DEAR LYELL,—Out of justice to Chambers I must trouble you with one line to say, as far as I am personally concerned in Glen Roy, he has made the amende honorable, and pleads guilty through inadvertency of taking my two lines of arguments and facts without acknowledgment.   He concluded by saying he "came to the same point by an independent course of inquiry, which in a small degree excuses this inadvertency."   His letter altogether shows a very good disposition, and says he is "much gratified with the *measured* approbation which you bestow, &c."   I am heartily glad I was able to say in truth that I thought he had

done good service in calling more attention to the subject of
the terraces. He protests it is unfair to call the sinking of
the sea his theory, for that he with care always speaks of mere
change of level, and this is quite true ; but the one section in
which he shows how he conceives the sea might sink is so
astonishing, that I believe it will with others, as with me,
more than counterbalance his previous caution. I hope that
you may think better of the book than I do.

Yours most truly,
C. DARWIN.

C. Darwin to J. D. Hooker.

October 6th, 1848.

. . . I have lately been trying to get up an agitation (but
I shall not succeed, and indeed doubt whether I have time
and strength to go on with it), against the practice of
Naturalists appending for perpetuity the name of the *first*
describer to species. I look at this as a direct premium to
hasty work, to *naming* instead of *describing*. A species ought
to have a name so well known that the addition of the author's
name would be superfluous, and a [piece] of empty vanity.*

* His contempt for the self-regarding spirit in a naturalist is illustrated by an anecdote, for which I am indebted to Rev. L. Blomefield. After speaking of my father's love of Entomology at Cambridge, Mr. Blomefield continues :—"He occasionally came over from Cambridge to my Vicarage at Swaffham Bulbeck, and we went out together to collect insects in the woods at Bottisham Hall, close at hand, or made longer excursions in the Fens. On one occasion he captured in a large bag net, with which he used vigorously to sweep the weeds and long grass, a rare coleopterous insect, one of the *Lepturidæ*, which I myself had never taken in Cambridgeshire. He was pleased with his capture, and of course carried it home in triumph. Some years afterwards, the voyage of the *Beagle* having been made in the interim, talking over old times with him, I reverted to this circumstance, and asked if he remembered it. 'Oh yes,' (he said,) 'I remember it well ; and I was selfish enough to keep the specimen, when you were collecting materials for a Fauna of Cambridgeshire, and for a local museum in the Philosophical Society.' He followed this up with some remarks on the pettiness of collectors, who aimed at nothing beyond filling their cabinets with rare things."

At present, it would not do to give mere specific names; but I think Zoologists might open the road to the omission, by referring to good systematic writers instead of to first describers. Botany, I fancy, has not suffered so much as Zoology from mere *naming;* the characters, fortunately, are more obscure. Have you ever thought on this point? Why should Naturalists append their own names to new species, when Mineralogists and Chemists do not do so to new substances? When you write to Falconer pray remember me affectionately to him. I grieve most sincerely to hear that he has been ill. My dear Hooker, God bless you, and fare you well.

Your sincere friend,
C. DARWIN.

*C. Darwin to Hugh Strickland.**

Down, Jan. 29th [1849].

. . . . . What a labour you have undertaken; I do *honour* your devoted zeal in the good cause of Natural Science. Do

---

* Hugh Edwin Strickland, M.A., F.R.S., was born 2nd of March, 1811, and educated at Rugby, under Arnold, and at Oriel College, Oxford. In 1835 and 1836 he travelled through Europe to the Levant with W. J. Hamilton, the geologist, wintering in Asia Minor. In 1841 he brought the subject of Natural History Nomenclature before the British Association, and prepared the Code of Rules for Zoological Nomenclature, now known by his name—the principles of which are very generally adopted. In 1843 he was one of the founders (if not the original projector) of the Ray Society. In 1845 he married the second daughter of Sir William Jardine, Bart. In 1850 he was appointed, in consequence of Buckland's illness, Deputy Reader in Geology at Oxford. His promising career was suddenly cut short on September 14, 1853, when, while geologizing in a railway cutting between Retford and Gainsborough, he was run over by a train and instantly killed. A memoir of him and a reprint of his principal contributions to journals was published by Sir William Jardine in 1858; but he was also the author of 'The Dodo and its Kindred' (1848); 'Bibliographia Zoologiæ' (the latter in conjunction with Louis Agassiz, and issued by the Ray Society); 'Ornithological Synonyms' (one volume only published, and that posthumously). A catalogue of his ornithological collection, given by his widow to the University of Cambridge, was compiled by Mr. Salvin, and published in 1882. (I am indebted to Prof. Newton for the above note).

you happen to have a *spare* copy of the Nomenclature rules published in the 'British Association Transactions?' if you have, and would give it me, I should be truly obliged, for I grudge buying the volume for it. I have found the rules very useful, it is quite a comfort to have something to rest on in the turbulent ocean of nomenclature (and am accordingly grateful to you), though I find it very difficult to obey always. Here is a case (and I think it should have been noticed in the rules), Coronula, Cineras and Otion, are names adopted by Cuvier, Lamarck, Owen, and almost *every* well-known writer, but I find that all three names were anticipated by a German : now I believe if I were to follow the strict rule of priority, more harm would be done than good, and more especially as I feel sure that the newly fished-up names would not be adopted.   I have almost made up my mind to reject the rule of priority in this case ; would you grudge the trouble to send me your opinion ?   I have been led of late to reflect much on the subject of naming, and I have come to a fixed opinion that the plan of the first describer's name, being appended for perpetuity to a species, has been the greatest curse to Natural History.   Some months since, I wrote out the enclosed badly drawn-up paper, thinking that perhaps I would agitate the subject; but the fit has passed, and I do not suppose I ever shall ; I send it you for the *chance* of your caring to see my notions.   I have been surprised to find in conversation that several naturalists were of nearly my way of thinking.   I feel sure as long as species-mongers have their vanity tickled by seeing their own names appended to a species, because they miserably described it in two or three lines, we shall have the same *vast* amount of bad work as at present, and which is enough to dishearten any man who is willing to work out any branch with care and time.   I find every genus of Cirripedia has half-a-dozen names, and not one careful description of any one species in any one genus. I do not believe that this would have been the case if each man knew that the memory of his own name depended on his doing his work well, and not upon merely appending a name with a few wretched lines indicating only a few prominent external characters. But I will not weary you with any longer tirade. Read my paper or *not*, just as you like, and return it whenever you please.

Yours most sincerely,

C. DARWIN.

Hugh Strickland to C. Darwin.

The Lodge, Tewkesbury, Jan. 31st, 1849.

. . . . I have next to notice your second objection—that retaining the name of the *first* describer in *perpetuum* along with that of the species, is a premium on hasty and careless work. This is quite a different question from that of the law of priority itself, and it never occurred to me before, though it seems highly probable that the general recognition of that law may produce such a result. We must try to counteract this evil in some other way.

The object of appending the name of a man to the name of a species is not to gratify the vanity of the man, but to indicate more precisely the species. Sometimes two men will, by accident, give the same name (independently) to two species of the same genus. More frequently a later author will misapply the specific name of an older one. Thus the *Helix putris* of Montague is not *H. putris* of Linnæus, though Montague supposed it to be so. In such a case we cannot define the species by *Helix putris* alone, but must append the name of the author whom we quote. But when a species has never borne but one name (as *Corvus frugilegus*), and no other species of Corvus has borne the same name, it is, of course, unnecessary to add the author's name. Yet even here I like the form *Corvus frugilegus, Linn.*, as it reminds us that this is one of the old species, long known, and to be found in the 'Systema

Naturæ,' &c.   I fear, therefore, that (at least until our nomenclature is more definitely settled) it will be impossible to
indicate species with scientific accuracy, without adding the
name of their first author.   You may, indeed, do it as you
propose, by saying *in Lam. An. Invert.*, *&c.*, but then this
would be incompatible with the law of priority, for where
Lamarck has violated that law, one cannot adopt his name.
It is, nevertheless, highly conducive to accurate indication to
append to the (oldest) specific name *one* good reference to a
standard work, especially to a *figure*, with an accompanying
synonym if necessary.   This method may be cumbrous, but
cumbrousness is a far less evil than uncertainty.

It, moreover, seems hardly possible to carry out the *priority*
principle without the historical aid afforded by appending the
author's name to the specific one.   If I, a *priority man*, called
a species *C. D.*, it implies that C. D. is the oldest name that
I know of ; but in order that you and others may judge of
the propriety of that name, you must ascertain when, and by
whom, the name was first coined.   Now, if to the specific
name C. D., I append the name A. B., of its first describer, I
at once furnish you with the clue to the dates when, and the
book in which, this description was given, and I thus assist
you in determining whether C. D. be really the oldest, and
therefore the correct, designation.

I do, however, admit that the priority principle (excellent
as it is) has a tendency, when the author's name is added, to
encourage vanity and slovenly work.   I think, however, that
much might be done to discourage those obscure and unsatisfactory definitions of which you so justly complain, by *writing
down* the practice.   Let the better disposed naturalists com-
bine to make a formal protest against all vague, loose, and
inadequate definitions of (supposed) new species.   Let a
committee (say of the British Association) be appointed to
prepare a sort of *Class List* of the various modern works in
which new species are described, arranged in order of merit.

The lowest class would contain the worst examples of the
kind, and their authors would thus be exposed to the obloquy
which they deserve, and be gibbeted *in terrorem* for the
edification of those who may come after.

I have thus candidly stated my views (I hope intelligibly)
of what seems best to be done in the present transitional and
dangerous state of systematic zoology. Innumerable labourers, many of them crotchety and half-educated, are
rushing into the field, and it depends, I think, on the present
generation whether the science is to descend to posterity a
chaotic mass, or possessed of some traces of law and organisa-
tion. If we could only get a congress of deputies from the
chief scientific bodies of Europe and America, something
might be done, but, as the case stands, I confess I do not
clearly see my way, beyond humbly endeavouring to reform
Number One.

Yours ever,

H. E. STRICKLAND.

C. Darwin to Hugh Strickland.

Down, Sunday [Feb. 4th, 1849].

MY DEAR STRICKLAND,—I am, in truth, *greatly* obliged to
you for your long, most interesting, and clear letter, and the
Report. I will consider your arguments, which are of the
greatest weight, but I confess I cannot yet bring myself to
reject very *well-known* names, not in *one* country, but over the
world, for obscure ones,—simply on the ground that I do not
believe I should be followed. Pray believe that I should
break the law of priority only in rare cases ; will you read the
enclosed (and return it), and tell me whether it does not
stagger you ? (N.B. I *promise* that I will not give you any
more trouble.) I want simple answers, and not for you to
waste your time in reasons ; I am curious for your answer in
regard to Balanus. I put the case of Otion, &c., to W.

Thompson, who is fierce for the law of priority, and he gave it up in such well-known names. I am in a perfect maze of doubt on nomenclature. In not one large genus of Cirripedia has *any one* species been correctly defined; it is pure guess-work (being guided by range and commonness and habits) to recognise any species: thus I can make out, from plates or descriptions, hardly any of the British sessile cirripedes. I cannot bear to give new names to all the species, and yet I shall perhaps do wrong to attach old names by little better than guess; I cannot at present tell the least which of two species all writers have meant by the common *Anatifera lævis*; I have, therefore, given that name to the one which is rather the commonest. Literally, not one species is properly defined; not one naturalist has ever taken the trouble to open the shell of any species to describe it scientifically, and yet all the genera have half-a-dozen synonyms. For *argument's* sake, suppose I do my work thoroughly well, any one who happens to have the original specimens named, I will say by Chenu, who has figured and named hundreds of species, will be able to upset all my names according to the law of priority (for he may maintain his descriptions are sufficient), do you think it advantageous to science that this should be done: I think not, and that convenience and high merit (here put as mere argument) had better come into some play. The subject is heart-breaking.

I hope you will occasionally turn in your mind my argument of the evil done by the "mihi" attached to specific names; I can most clearly see the *excessive* evil it has caused; in mineralogy I have myself found there is no rage to merely name; a person does not take up the subject without he intends to work it out, as he knows that his *only* claim to merit rests on his work being ably done, and has no relation whatever to *naming*. I give up one point, and grant that reference to first describer's name should be given in all systematic works, but I think something would be gained if a reference was given without the author's name being actually appended as part of the binomial name, and I think, except in systematic works, a reference, such as I propose, would damp vanity much.   I think a very wrong spirit runs through all Natural History, as if some merit was due to a man for merely naming and defining a species; I think scarcely any, or none, is due; if he works out *minutely* and anatomically any one species, or systematically a whole group, credit is due, but I must think the mere defining a species is nothing, and that no *injustice* is done him if it be overlooked, though a great inconvenience to Natural History is thus caused.   I do not think more credit is due to a man for defining a species, than to a carpenter for making a box.   But I am foolish and rabid against species-mongers, or rather against their vanity; it is useful and necessary work which must be done; but they act as if they had actually made the species, and it was their own property.

I use Agassiz's nomenclator; at least two-thirds of the dates in the Cirripedia are grossly wrong.

I shall do what I can in fossil Cirripedia, and should be very grateful for specimens; but I do not believe that species (and hardly genera) can be defined by single valves; as in every recent species yet examined their forms vary greatly: to describe a species by valves alone, is the same as to describe a crab from *small* portions of its carapace alone, these portions being highly variable, and not, as in Crustacea, modelled over viscera.   I sincerely apologise for the trouble which I have given you, but indeed I will give no more.

Yours most sincerely,

C. DARWIN.

P.S.—In conversation I found Owen and Andrew Smith much inclined to throw over the practice of attaching authors' names; I believe if I agitated I could get a large party to join. W. Thompson agreed some way with me, but was not prepared to go nearly as far as I am.

C. Darwin to Hugh Strickland.

Down, Feb. 10th [1849].

MY DEAR STRICKLAND,—I have again to thank you cordially for your letter. Your remarks shall fructify to some extent, and I will try to be more faithful to rigid virtue and priority; but as for calling Balanus "Lepas" (which I did not think of), I cannot do it, my pen won't write it—it is *impossible*. I have great hopes some of my difficulties will disappear, owing to wrong dates in Agassiz, and to my having to run several genera into one, for I have as yet gone, in but few cases, to original sources. With respect to adopting my own notions in my Cirripedia book, I should not like to do so without I found others approved, and in some public way— nor, indeed, is it well adapted, as I can never recognise a species without I have the original specimen, which, fortunately, I have in many cases in the British Museum. Thus far I mean to adopt my notion, as never putting mihi or "Darwin" after my own species, and in the anatomical text giving no authors' names at all, as the systematic Part will serve for those who want to know the History of a species as far as I can imperfectly work it out. . . . . .

C. Darwin to J. D. Hooker.

[The Lodge, Malvern,
March 28th, 1849.]

MY DEAR HOOKER,—Your letter of the 13th of October has remained unanswered till this day! What an ungrateful return for a letter which interested me so much, and which contained so much and curious information. But I have had a bad winter.

On the 13th of November, my poor dear father died, and no one who did not know him would believe that a man above eighty-three years old could have retained so tender and affectionate a disposition, with all his sagacity unclouded to the last. I was at the time so unwell, that I was unable to

travel, which added to my misery. Indeed, all this winter I
have been bad enough . . . and my nervous system began
to be affected, so that my hands trembled, and head was
often swimming. I was not able to do anything one day out
of three, and was altogether too dispirited to write to you,
or to do anything but what I was compelled. I thought I
was rapidly going the way of all flesh. Having heard, acci-
dentally, of two persons who had received much benefit from
the water-cure, I got Dr. Gully's book, and made further
enquiries, and at last started here, with wife, children, and
all our servants. We have taken a house for two months,
and have been here a fortnight. I am already a little
stronger . . . Dr. Gully feels pretty sure he can do me good,
which most certainly the regular doctors could not. . . . . . .
I feel certain that the water-cure is no quackery.

How I shall enjoy getting back to Down with renovated
health, if such is to be my good fortune, and resuming the
beloved Barnacles. Now I hope that you will forgive me for
my negligence in not having sooner answered your letter. I
was uncommonly interested by the sketch you give of your
intended grand expedition, from which I suppose you will
soon be returning. How earnestly I hope that it may prove
in every way successful. . . .

[When my father was at the Water-cure Establishment at
Malvern he was brought into contact with clairvoyance, of
which he writes in the following extract from a letter to Fox,
September, 1850.

" You speak about Homœopathy, which is a subject which
makes me more wrath, even than does Clairvoyance. Clairvoyance so transcends belief, that one's ordinary faculties are
put out of the question, but in homœopathy common sense
and common observation come into play, and both these must
go to the dogs, if the infinitesimal doses have any effect whatever. How true is a remark I saw the other day by Quetelet,

in respect to evidence of curative processes, viz. that no one knows in disease what is the simple result of nothing being done, as a standard with which to compare homœopathy, and all other such things. It is a sad flaw, I cannot but think, in my beloved Dr. Gully, that he believes in everything. When Miss —— was very ill, he had a clairvoyant girl to report on internal changes, a mesmerist to put her to sleep—an homœopathist, viz. Dr. ——, and himself as hydropathist! and the girl recovered."

A passage out of an earlier letter to Fox (December, 1844) shows that he was equally sceptical on the subject of mesmerism : "With respect to mesmerism, the whole country resounds with wonderful facts or tales . . . I have just heard of a child, three or four years old (whose parents and self I well knew), mesmerised by his father, which is the first fact which has staggered me. I shall not believe fully till I see or hear from good evidence of animals (as has been stated is possible) not drugged, being put to stupor ; of course the impossibility would not prove mesmerism false ; but it is the only clear *experimentum crucis*, and I am astonished it has not been systematically tried. If mesmerism was investigated, like a science, this could not have been left till the present day to be *done satisfactorily*, as it has been I believe left. Keep some cats yourself, and do get some mesmeriser to attempt it. One man told me he had succeeded, but his experiments were most vague, as was likely from a man who said cats were more easily done than other animals, because they were so electrical !"]

C. Darwin to C. Lyell.

Down, December 4th [1849].

MY DEAR LYELL,—This letter requires no answer, and I write from exuberance of vanity. Dana has sent me the Geology of the United States Expedition, and I have just read the Coral part. To begin with a modest speech, *I am astonished at my own accuracy!!* If I were to rewrite now my Coral book there is hardly a sentence I should have to alter, except that I ought to have attributed more effect to recent volcanic action in checking growth of coral. When I say all this I ought to add that the *consequences* of the theory on areas of subsidence are treated in a separate chapter to which I have not come, and in this, I suspect, we shall differ more. Dana talks of agreeing with my theory *in most points ;* I can find out not one in which he differs. Considering how infinitely more he saw of Coral Reefs than I did, this is wonderfully satisfactory to me. He treats me most courteously. There now, my vanity is pretty well satisfied. . . .

C. Darwin to J. D. Hooker.

Malvern, April 9th, 1849.

MY DEAR HOOKER,—The very next morning after posting my last letter (I think on 23rd of March), I received your two interesting gossipaceous and geological letters ; and the latter I have since exchanged with Lyell for his. I will write higglety-pigglety just as subjects occur. I saw the Review in the ' Athenæum,' it was written in an ill-natured spirit ; but the whole virus consisted in saying that there was not novelty enough in your remarks for publication. No one, nowadays, cares for reviews. I may just mention that my Journal got some *real good* abuse, " presumption," &c.—ended with saying that the volume appeared " made up of the scraps and rubbish of the author's portfolio." I most truly enter into what you say, and quite believe you that you care only for the review with respect to your father ; and that this *alone* would make you like to see extracts from your letters more properly noticed in this same periodical. I have considered to the very best of my judgment whether any portion of your present letters are adapted for the ' Athenæum ' (in which I have no

interest ; the beasts not having even *noticed* my three geological volumes which I had sent to them), and I have come
to the conclusion it is better not to send them.   I feel sure,
considering all the circumstances, that without you took pains
and wrote *with care*, a condensed and finished sketch of some
striking feature in your travels, it is better not to send
anything.   These two letters are, moreover, rather too geological for the 'Athenæum,' and almost require woodcuts.   On
the other hand, there are hardly enough details for a communication to the Geological Society.   I have not the *smallest
doubt* that your facts are of the highest interest with regard to
glacial action in the Himalaya ; but it struck both Lyell and
myself that your evidence ought to have been given more
distinctly. . . .

I have written so lately that I have nothing to say about
myself; my health prevented me going on with a crusade
against "mihi." and "nobis," of which you warn me of the
dangers.   I showed my paper to three or four Naturalists, and
they all agreed with me to a certain extent : with health and
vigour, I would not have shown a white feather, [and] with
aid of half-a-dozen really good Naturalists, I believe something
might have been done against the miserable and degrading
passion of mere species naming.   In your letter you wonder
what "Ornamental Poultry" has to do with Barnacles ; but
do not flatter yourself that I shall not yet live to finish the
Barnacles, and then make a fool of myself on the subject of
species, under which head ornamental Poultry are very
interesting. . . .

C. Darwin to C. Lyell.

The Lodge, Malvern [June, 1849].

. . . I have got your book,* and have read all the first and a
small part of the second volume (reading is the hardest work

* 'A Second Visit to the United States.'

allowed here), and greatly I have been interested by it. It makes me long to be a Yankee. E. desires me to say that she quite "gloated" over the truth of your remarks on religious progress .... I delight to think how you will disgust some of the bigots and educational dons. As yet there has not been *much* Geology or Natural History, for which I hope you feel a little ashamed. Your remarks on all social subjects strike me as worthy of the author of the 'Principles.' And yet (I know it is prejudice and pride) if I had written the Principles, I never would have written any travels; but I believe I am more jealous about the honour and glory of the Principles than you are yourself. . . .

C. Darwin to C. Lyell.

September 14th, 1849.

... I go on with my aqueous processes, and very steadily but slowly gain health and strength. Against all rules, I dined at Chevening with Lord Mahon, who did me the great honour of calling on me, and how he heard of me I can't guess. I was charmed with Lady Mahon, and any one might have been proud at the pieces of agreeableness which came from her beautiful lips with respect to you. I like old Lord Stanhope very much; though he abused Geology and Zoology heartily. "To suppose that the Omnipotent God made a world, found it a failure, and broke it up, and then made it again, and again broke it up, as the Geologists say, is all fiddle faddle." Describing Species of birds and shells, &c., is all fiddle faddle. . . .

I am heartily glad we shall meet at Birmingham, as I trust we shall, if my health will but keep up. I work now every day at the Cirripedia for 2½ hours, and so get on a little, but very slowly. I sometimes, after being a whole week employed and having described perhaps only two species, agree mentally with Lord Stanhope, that it is all fiddle faddle; however, the other day I got the curious case of a unisexual, instead of hermaphrodite cirripede, in which the female had the common cirripedial character, and in two valves of her shell had two little pockets, in *each* of which she kept a little husband ; I do not know of any other case where a female invariably has two husbands. I have one still odder fact, common to several species, namely, that though they are hermaphrodite, they have small additional, or as I shall call them, complemental males, one specimen itself hermaphrodite had no less than *seven*, of these complemental males attached to it. Truly the schemes and wonders of Nature are illimitable. But I am running on as badly about my cirripedia as about Geology ; it makes me groan to think that probably I shall never again have the exquisite pleasure of making out some new district, of evolving geological light out of some troubled dark region. So I must make the best of my Cirripedia. . . .

C. Darwin to J. D. Hooker.

Down, October 12th, 1849.

. . . By the way, one of the pleasantest parts of the British Association was my journey down to Birmingham with Mrs. Sabine, Mrs. Reeve, and the Colonel; also Col. Sykes and Porter. Mrs. Sabine and myself agreed wonderfully on many points, and in none more sincerely than about you. We spoke about your letters from the Erebus ; and she quite agreed with me, that you and the *author* * of the description of the cattle hunting in the Falklands, would have made a capital book together! A very nice woman she is, and so is her sharp and sagacious mother. . . . Birmingham was very flat compared to Oxford, though I had my wife with me. We saw a good deal of the Lyells and Horners and Robinsons (the President); but the place was dismal, and

---

* Sir J. Hooker wrote the spirited description of cattle hunting in Sir J. Ross's 'Voyage of Discovery in the Southern Regions,' 1847, vol. ii. p. 245.

I was prevented, by being unwell, from going to Warwick, though that, *i.e.* the party, by all accounts, was wonderfully inferior to Blenheim, not to say anything of that heavenly day at Dropmore.   One gets weary of all the spouting. . . .

You ask about my cold-water cure; I am going on very well, and am certainly a little better every month, my nights mend much slower than my days.   I have built a douche, and am to go on through all the winter, frost or no frost. My treatment now is lamp five times per week, and shallow bath for five minutes afterwards; douche daily for five minutes, and dripping sheet daily. The treatment is wonderfully tonic, and I have had more better consecutive days this month than on any previous ones. . . . I am allowed to work now two and a half hours daily, and I find it as much as I can do; for the cold-water cure, together with three short walks, is curiously exhausting; and I am actually *forced* to go to bed at eight o'clock completely tired. I steadily gain in weight, and eat immensely, and am never oppressed with my food. I have lost the involuntary twitching of the muscle, and all the fainting feelings, &c.—black spots before eyes, &c. Dr. Gully thinks he shall quite cure me in six or nine months more.

The greatest bore, which I find in the water-cure, is the having been compelled to give up all reading, except the news-papers; for my daily two and a half hours at the Barnacles is fully as much as I can do of anything which occupies the mind; I am consequently terribly behind in all scientific books. I have of late been at work at mere species de-scribing, which is much more difficult than I expected, and has much the same sort of interest as a puzzle has; but I confess I often feel wearied with the work, and cannot help sometimes asking myself what is the good of spending a week or fortnight in ascertaining that certain just perceptible dif-ferences blend together and constitute varieties and not species. As long as I am on anatomy I never feel myself in that disgusting, horrid, *cui bono*, inquiring, humour. What miserable work, again, it is searching for priority of names. I have just finished two species, which possess seven generic, and twenty-four specific names! My chief comfort is, that the work must be sometime done, and I may as well do it, as any one else.

I have given up my agitation against *mihi* and *nobis;* my paper is too long to send to you, so you must see it, if you care to do so, on your return. By-the-way, you say in your letter that you care more for my species work than for the Barnacles; now this is too bad of you, for I declare your decided approval of my plain Barnacle work over theoretic species work, had very great influence in deciding me to go on with the former, and defer my species paper. . . .

[The following letter refers to the death of his little daughter, which took place at Malvern on April 24, 1851 :]

C. Darwin to W. D. Fox.

Down, April 29th [1851].

MY DEAR FOX,—I do not suppose you will have heard of our bitter and cruel loss. Poor dear little Annie, when going on very well at Malvern, was taken with a vomiting attack, which was at first thought of the smallest importance ; but it rapidly assumed the form of a low and dreadful fever, which carried her off in ten days. Thank God, she suffered hardly at all, and expired as tranquilly as a little angel. Our only consolation is that she passed a short, though joyous life. She was my favourite child ; her cordiality, openness, buoyant joyousness and strong affections made her most loveable. Poor dear little soul. Well, it is all over. . . .

C. Darwin to W. D. Fox.

Down, March 7th [1852].

MY DEAR FOX,—It is indeed an age since we have had any communication, and very glad I was to receive your note.

Our long silence occurred to me a few weeks since, and I had then thought of writing, but was idle. I congratulate and condole with you on your *tenth* child; but please to observe when I have a tenth, send only condolences to me. We have now seven children, all well, thank God, as well as their mother; of these seven, five are boys; and my father used to say that it was certain that a boy gave as much trouble as three girls; so that *bonâ fide* we have seventeen children. It makes me sick whenever I think of professions; all seem hopelessly bad, and as yet I cannot see a ray of light. I should very much like to talk over this (by the way, my three bug-bears are Californian and Australian gold, beggaring me by making my money on mortgage worth nothing; the French coming by the Westerham and Sevenoaks roads, and there-fore enclosing Down; and thirdly, professions for my boys), and I should like to talk about education, on which you ask me what we are doing. No one can more truly despise the old stereotyped stupid classical education than I do; but yet I have not had courage to break through the trammels. After many doubts we have just sent our eldest boy to Rugby, where for his age he has been very well placed. . . . I honour, admire, and envy you for educating your boys at home. What on earth shall you do with your boys? Towards the end of this month we go to see W. at Rugby, and thence for five or six days to Susan * at Shrewsbury; I then return home to look after the babies, and E. goes to F. Wedgwood's of Etruria for a week. Very many thanks for your most kind and large invitation to Delamere, but I fear we can hardly compass it. I dread going anywhere, on account of my stomach so easily failing under any excite-ment. I rarely even now go to London; not that I am at all worse, perhaps rather better, and lead a very comfortable life with my three hours of daily work, but it is the life of a hermit. My nights are *always* bad, and that stops my

* His sister.

becoming vigorous. You ask about water-cure. I take at intervals of two or three months, five or six weeks of *moderately* severe treatment, and always with good effect. Do you come here, I pray and beg whenever you can find time ; you cannot tell how much pleasure it would give me and E. I have finished the 1st vol. for the Ray Society of Peduncula-ted Cirripedes, which, as I think you are a member, you will soon get. Read what I describe on the sexes of Ibla and Scalpellum. I am now at work on the Sessile Cirripedes, and am wonderfully tired of my job : a man to be a systematic naturalist ought to work at least eight hours per day. You saw through me, when you said that I must have wished to have seen the effects of the [word illegible] Debacle, for I was saying a week ago to E., that had I been as I was in old days, I would have been certainly off that hour. You ask after Erasmus ; he is much as usual, and constantly more or less unwell. Susan * is much better, and very flourishing and happy. Catherine * is at Rome, and has enjoyed it in a degree that is quite astonishing to my old dry bones. And now I think I have told you enough, and more than enough about the house of Darwin ; so my dear old friend, farewell. What pleasant times we had in drinking coffee in your rooms at Christ's College, and think of the glories of Crux major.† Ah, in those days there were no professions for sons, no ill-health to fear for them, no Californian gold, no French invasions. How paramount the future is to the present when one is surrounded by children. My dread is hereditary ill-health. Even death is better for them.

My dear Fox, your sincere friend,
C. DARWIN.

P.S.—Susan has lately been working in a way which I think truly heroic about the scandalous violation of the Act against children climbing chimneys. We have set up a

---

* His sisters.   † The beetle *Panagæus crux major.*

little Society in Shrewsbury to prosecute those who break the law. It is all Susan's doing. She has had very nice letters from Lord Shaftesbury and the Duke of Sutherland, but the brutal Shropshire squires are as hard as stones to move. The Act out of London seems most commonly violated. It makes one shudder to fancy one of one's own children at seven years old being forced up a chimney—to say nothing of the consequent loathsome disease and ulcerated limbs, and utter moral degradation. If you think strongly on this subject, do make some enquiries; add to your many good works, this other one, and try to stir up the magistrates. There are several people making a stir in different parts of England on this subject. It is not very likely that you would wish for such, but I could send you some essays and information if you so liked, either for yourself or to give away.

C. Darwin to W. D. Fox.

Down [October 24th, 1852].

My dear Fox,—I received your long and most welcome letter this morning, and will answer it this evening, as I shall be very busy with an artist, drawing Cirripedia, and much overworked for the next fortnight. But first you deserve to be well abused—and pray consider yourself well abused— for thinking or writing that I could for one minute be bored by any amount of detail about yourself and belongings. It is just what I like hearing; believe me that I often think of old days spent with you, and sometimes can hardly believe what a jolly careless individual one was in those old days. A bright autumn evening often brings to mind some shooting excursion from Osmaston. I do indeed regret that we live so far off each other, and that I am so little locomotive. I have been un- usually well of late (no water-cure), but I do not find that I can stand any change better than formerly. . . The other day I went to London and back, and the fatigue, though so trifling, brought on my bad form of vomiting. I grieve to hear that your chest has been ailing, and most sincerely do I hope that it is only the muscles; how frequently the voice fails with the clergy. I can well understand your reluctance to break up your large and happy party and go abroad; but your life is very valuable, so you ought to be very cautious in good time. You ask about all of us, now five boys (oh! the professions; oh! the gold; and oh! the French—these three oh's all rank as dreadful bugbears) and two girls . . . but another and the worst of my bugbears is hereditary weakness. All my sisters are well except Mrs. Parker, who is much out of health; and so is Erasmus at his poor average: he has lately moved into Queen Anne Street. I had heard of the intended marriage * of your sister Frances. I believe I have seen her since, but my memory takes me back some twenty-five years, when she was lying down. I remember well the delightful expression of her countenance. I most sincerely wish her all happiness.

I see I have not answered half your queries. We like very well all that we have seen and heard of Rugby, and have never repented of sending [W.] there. I feel sure schools have greatly improved since our days; but I hate schools and the whole system of breaking through the affections of the family by separating the boys so early in life; but I see no help, and dare not run the risk of a youth being exposed to the temptations of the world without having undergone the milder ordeal of a great school.

I see you even ask after our pears. We have had lots of Beurrées d'Aremberg, Winter Nelis, Marie Louise, and "Ne plus Ultra," but all off the wall; the standard dwarfs have borne a few, but I have no room for more trees, so their names would be useless to me. You really must make a holiday and pay us a visit sometime; nowhere could you be more heartily welcome. I am at work at the second volume

* To the Rev. J. Hughes.

of the Cirripedia, of which creatures I am wonderfully tired. I hate a Barnacle as no man ever did before, not even a sailor in a slow-sailing ship. My first volume is out ; the only part worth looking at is on the sexes of Ibla and Scalpellum. I hope by next summer to have done with my tedious work. Farewell,—do come whenever you can possibly manage it.

I cannot but hope that the carbuncle may possibly do you good ; I have heard of all sorts of weaknesses disappearing after a carbuncle. I suppose the pain is dreadful. I agree most entirely, what a blessed discovery is chloroform. When one thinks of one's children, it makes quite a little difference in one's happiness. The other day I had five grinders (two by the elevator) out at a sitting under this wonderful substance, and felt hardly anything.

My dear old friend, yours very affectionately,

CHARLES DARWIN.

C. Darwin to W. D. Fox.

Down, January 29th [1853].

MY DEAR FOX,—Your last account some months ago was so little satisfactory that I have often been thinking of you, and should be really obliged if you would give me a few lines, and tell me how your voice and chest are. I most sincerely hope that your report will be good. . . . Our second lad has a strong mechanical turn, and we think of making him an engineer. I shall try and find out for him some less classical school, perhaps Bruce Castle. I certainly should like to see more diversity in education than there is in any ordinary school—no exercising of the observing or reasoning faculties, no general knowledge acquired—I must think it a wretched system. On the other hand, a boy who has learnt to stick at Latin and conquer its difficulties, ought to be able to stick at any labour. I should always be glad to hear anything about schools or education from you. I am at my old, never-ending subject, but trust I shall really go to press in a few months with my second volume on Cirripedes. I have been much pleased by finding some odd facts in my first volume believed by Owen and a few others, whose good opinion I regard as final. . . . Do write pretty soon, and tell me all you can about yourself and family ; and I trust your report of yourself may be much better than your last.

. . . I have been very little in London of late, and have not seen Lyell since his return from America ; how lucky he was to exhume with his own hand parts of three skeletons of reptiles out of the *Carboniferous* strata, and out of the inside of a fossil tree, which had been hollow within.

Farewell, my dear Fox, yours affectionately,

CHARLES DARWIN.

C. Darwin to W. D. Fox.

13 Sea Houses, Eastbourne,
July [15th ? 1853].

MY DEAR FOX,—Here we are in a state of profound idleness, which to me is a luxury ; and we should all, I believe, have been in a state of high enjoyment, had it not been for the detestable cold gales and much rain, which always gives much *ennui* to children away from their homes. I received your letter of 13th June, when working like a slave with Mr. Sowerby at drawing for my second volume, and so put off answering it till when I knew I should be at leisure. I was extremely glad to get your letter. I had intended a couple of months ago sending you a savage or supplicating jobation to know how you were, when I met Sir P. Egerton, who told me you were well, and, as usual, expressed his admiration of your doings, especially your farming, and the number of animals, including children, which you kept on your land. Eleven children, ave Maria ! it is a serious look-out for you. Indeed, I look at my five boys as something awful, and hate the very thoughts of professions, &c. If one could insure moderate health for them it would not signify so much, for I cannot but hope, with the enormous emigration, professions will some-what improve. But my bugbear is hereditary weakness. I particularly like to hear all that you can say about education, and you deserve to be scolded for saying "you did not mean to *torment* me with a long yarn." You ask about Rugby. I like it very well, on the same principle as my neighbour, Sir J. Lubbock, likes Eton, viz., that it is not worse than any other school; the expense, *with all, &c., &c.,* including some clothes, travelling expences, &c., is from £110 to £120 per annum. I do not think schools are so wicked as they were, and far more industrious. The boys, I think, live too secluded in their separate studies; and I doubt whether they will get so much knowledge of character as boys used to do; and this, in my opinion, is the *one* good of public schools over small schools. I should think the only superiority of a small school over home was forced regularity in their work, which your boys perhaps get at your home, but which I do not believe my boys would get at my home. Otherwise, it is quite lamentable sending boys so early in life from their home.

. . . To return to schools. My main objection to them, as places of education, is the enormous proportion of time spent over classics. I fancy (though perhaps it is only fancy) that I can perceive the ill and contracting effect on my eldest boy's mind, in checking interest in anything in which reasoning and observation come into play. Mere memory seems to be worked. I shall certainly look out for some school with more diversified studies for my younger boys. I was talking lately to the Dean of Hereford, who takes most strongly this view; and he tells me that there is a school at Hereford commencing on this plan; and that Dr. Kennedy at Shrewsbury is going to begin vigorously to modify that school. . . . .

I am *extremely* glad to hear that you approved of my cirri-pedial volume. I have spent an almost ridiculous amount of labour on the subject, and certainly would never have under-

taken it had I foreseen what a job it was. I hope to have
finished by the end of the year. Do write again before a very
long time ; it is a real pleasure to me to hear from you.
Farewell, with my wife's kindest remembrances to yourself
and Mrs. Fox.

My dear old friend, yours affectionately,

C. DARWIN.

C. Darwin to W. D. Fox.

Down, August 10th [1853].

MY DEAR FOX,—I thank you sincerely for writing to me so
soon after your most heavy misfortunes. Your letter affected
me much. We both most truly sympathise with you and
Mrs. Fox. We too lost, as you may remember, not so very
long ago, a most dear child, of whom I can hardly yet bear
to think tranquilly ; yet, as you must know from your own
most painful experience, time softens and deadens, in a
manner truly wonderful, one's feelings and regrets. At first
it is indeed bitter. I can only hope that your health and
that of poor Mrs. Fox may be preserved, and that time may
do its work softly, and bring you all together, once again, as
the happy family, which, as I can well believe, you so
lately formed.

My dear Fox, your affectionate friend,

CHARLES DARWIN.

[The following letter refers to the Royal Society's Medal,
which was awarded to him in November, 1853 :]

C. Darwin to J. D. Hooker.

Down, November 5th [1853].

MY DEAR HOOKER,—Amongst my letters received this
morning, I opened first one from Colonel Sabine ; the contents
certainly surprised me very much, but, though the letter was

a *very kind one*, somehow, I cared very little indeed for the announcement it contained. I then opened yours, and such is the effect of warmth, friendship, and kindness from one that is loved, that the very same fact, told as you told it, made me glow with pleasure till my very heart throbbed. Believe me, I shall not soon forget the pleasure of your letter. Such hearty, affectionate sympathy is worth more than all the medals that ever were or will be coined. Again, my dear Hooker, I thank you. I hope Lindley * will never hear that he was a competitor against me ; for really it is almost *ridiculous* (of course you would never repeat that I said this, for it would be thought by others, though not, I believe, by you, to be affectation) his not having the medal long before me ; I must feel *sure* that you did quite right to propose him ; and what a good, dear, kind fellow you are, nevertheless, to rejoice in this honour being bestowed on me.

What *pleasure* I have felt on the occasion, I owe almost entirely to you.

Farewell, my dear Hooker, yours affectionately,

C. DARWIN.

---

* John Lindley (b. 1799, d. 1865) was the son of a nurseryman near Norwich, through whose failure in business he was thrown at the age of twenty on his own resources. He was befriended by Sir W. Hooker, and employed as assistant librarian by Sir J. Banks. He seems to have had enormous capacity of work, and is said to have translated Richard's 'Analyse du Fruit' at one sitting of two days and three nights. He became Assistant-Secretary to the Horticultural Society, and in 1829 was appointed Professor of Botany at University College, a post which he held for upwards of thirty years. His writings are numerous : the best known being perhaps his 'Vegetable Kingdom,' published in 1846. His influence in helping to introduce the natural system of classification was considerable, and he brought " all the weight of his teaching and all the force of his controversial powers to support it," as against the Linnean system universally taught in the earlier part of his career. Sachs points out (Geschichte der Botanik, 1875, p. 161), that though Lindley adopted in the main a sound classification of plants, he only did so by abandoning his own theoretical principle that the physiological importance of an organ is a measure of its classificatory value.

P.S.—You may believe what a surprise it was, for I had never heard that the medals could be given except for papers in the ' Transactions.' All this will make me work with better heart at finishing the second volume.

C. Darwin to C. Lyell.

Down, February 18th [1854].

MY DEAR LYELL,—I should have written before, had it not seemed doubtful whether you would go on to Teneriffe, but now I am extremely glad to hear your further progress is certain; not that I have much of any sort to say, as you may well believe when you hear that I have only once been in London since you started. I was particularly glad to see, two days since, your letter to Mr. Horner, with its geological news; how fortunate for you that your knees are recovered. I am astonished at what you say of the beauty, though I had fancied it great. It really makes me quite envious to think of your clambering up and down those steep valleys. And what a pleasant party on your return from your expeditions. I often think of the delight which I felt when examining volcanic islands, and I can remember even particular rocks which I struck, and the smell of the hot, black, scoriaceous cliffs; but of those *hot* smells you do not seem to have had much. I do quite envy you. How I should like to be with you, and speculate on the deep and narrow valleys.

How very singular the fact is which you mention about the inclination of the strata being greater round the circumference than in the middle of the island; do you suppose the elevation has had the form of a flat dome. I remember in the Cordillera being *often* struck with the greater abruptness of the strata in the *low extreme* outermost ranges, compared with the great mass of inner mountains. I dare say you will have thought of measuring exactly the width of any dikes at the top and bottom of any great cliff (which was done by

Mr. Searle [?] at St. Helena), for it has often struck me as *very
odd* that the cracks did not die out *oftener* upwards. I can
think of hardly any news to tell you, as I have seen no one
since being in London, when I was delighted to see Forbes
looking so well, quite big and burly. I saw at the Museum
some of the surprisingly rich gold ore from North Wales.
Ramsay also told me that he has lately turned a good deal of
New Red Sandstone into Permian, together with the Laby-
rinthodon. No doubt you see newspapers, and know that
E. de Beaumont is perpetual Secretary, and will, I suppose,
be more powerful than ever ; and Le Verrier has Arago's
place in the Observatory. There was a meeting lately at the
Geological Society, at which Prestwich (judging from what
R. Jones told me) brought forward your exact theory, viz.
that the whole red clay and flints over the chalk plateau
hereabouts is the residuum from the slow dissolution of the
chalk !

As regards ourselves, we have no news, and are all well.
The Hookers, sometime ago, stayed a fortnight with us, and,
to our extreme delight, Henslow came down, and was most
quiet and comfortable here. It does one good to see so com-
posed, benevolent, and intellectual a countenance. There
have been great fears that his heart is affected ; but, I hope to
God, without foundation. Hooker's book * is out, and *most
beautifully* got up. He has honoured me beyond measure by
dedicating it to me ! As for myself, I am got to the page 112
of the Barnacles, and that is the sum total of my history.
By-the-way, as you care so much about North America, I
may mention that I had a long letter from a ship-mate in
Australia, who says the Colony is getting decidedly repub-
lican from the influx of Americans, and that all the great and
novel schemes for working the gold are planned and executed
by these men. What a go-a-head nation it is ! Give my
kindest remembrances to Lady Lyell, and to Mrs. Bunbury,

* Sir J. Hooker's ' Himalayan Journal.'

and to Bunbury.   I most heartily wish that the Canaries may
be ten times as interesting as Madeira, and that everything
may go on most prosperously with your whole party.

   My dear Lyell,

     Yours most truly and affectionately,

         C. DARWIN.

    C. Darwin to J. D. Hooker

        Down, March 1st [1854].

MY DEAR HOOKER,—I finished yesterday evening the first
volume, and I very sincerely congratulate you on having
produced a *first-class* book *—a book which certainly will last.
I cannot doubt that it will take its place as a standard, not so
much because it contains real solid matter, but that it gives
a picture of the whole country.   One can feel that one
has seen it (and desperately uncomfortable I felt in going
over some of the bridges and steep slopes), and one *realises*
all the great Physical features.   You have in truth reason to
be proud ; consider how few travellers there have been with
a profound knowledge of one subject, and who could in
addition make a map (which, by-the-way, is one of the most
distinct ones I ever looked at, wherefore blessings alight on
your head), and study geology and meteorology !   I thought
I knew you very well, but I had not the least idea that
your Travels were your hobby ; but I am heartily glad of it,
for I feel sure that the time will never come when you and
Mrs. Hooker will not be proud to look back at the labour
bestowed on these beautiful volumes.

 Your letter, received this morning, has interested me
extremely, and I thank you sincerely for telling me your old
thoughts and aspirations.   All that you say makes me
even more deeply gratified by the Dedication ; but you,
bad man, do you remember asking me how I thought Lyell
would like the work to be dedicated to him ?   I remember

    * ' Himalayan Journal.'

how strongly I answered, and I presume you wanted to know what I should feel; whoever would have dreamed of your being so crafty? I am glad you have shown a little bit of ambition about your Journal, for you must know that I have often abused you for not caring more about fame, though, at the same time, I must confess, I have envied and honoured you for being so free (too free, as I have always thought) of this "last infirmity of, &c." Do not say, "there never was a past hitherto to me—the phantom was always in view," for you will soon find other phantoms in view. How well I know this feeling, and did formerly still more vividly; but I think my stomach has much deadened my former pure enthusiasm for science and knowledge.

I am writing an unconscionably long letter, but I must return to the Journals, about which I have hardly said anything in detail. Imprimis, the illustrations and maps appear to me the best I have ever seen; the style seems to me everywhere perfectly clear (how rare a virtue), and some passages really eloquent. How excellently you have described the upper valleys, and how detestable their climate; I felt quite anxious on the slopes of Kinchin that dreadful snowy night. Nothing has astonished me more than your physical strength; and all those devilish bridges! Well, thank goodness! it is not *very* likely that I shall ever go to the Himalaya. Much in a scientific point of view has interested me, especially all about those wonderful moraines. I certainly think I quite realise the valleys, more vividly perhaps from having seen the valleys of Tahiti. I cannot doubt that the Himalaya owe almost all their contour to running water, and that they have been subjected to such action longer than any mountains (as yet described) in the world. What a contrast with the Andes!

Perhaps you would like to hear the very little that I can say *per contra*, and this only applied to the beginning, in which (as it struck me) there was not *flow* enough till you get to

Mirzapore on the Ganges (but the Thugs were *most* interesting),
where the stream seemed to carry you on more equably with
longer sentences and longer facts and discussions, &c. In
another edition (and I am delighted to hear that Murray has
sold all off), I would consider whether this part could not be
condensed. Even if the meteorology was put in foot-notes, I
think it would be an improvement. All the world is against
me, but it makes me very unhappy to see the Latin names all
in Italics, and all mingled with English names in Roman type ;
but I must bear this burden, for all men of Science seem to
think it would corrupt the Latin to dress it up in the same
type as poor old English. Well, I am very proud of *my* book ;
but there is one bore, that I do not much like asking people
whether they have seen it, and how they like it, for I feel so
much identified with it, that such questions become rather
personal. Hence, I cannot tell you the opinion of others.
You will have seen a fairly good review in the ' Athenæum.'

What capital news from Tasmania : it really is a very
remarkable and creditable fact to the Colony.* I am always
building veritable castles in the air about emigrating, and
Tasmania has been my head-quarters of late ; so that I feel
very proud of my adopted country : it is really a very singular
and delightful fact, contrasted with the slight appreciation of
science in the old country. I thank you heartily for your
letter this morning, and for all the gratification your Dedication has given me ; I could not help thinking how much
——- would despise you for not having dedicated it to some
great man, who would have done you and it some good in the
eyes of the world. Ah, my dear Hooker, you were very soft
on this head, and justify what I say about not caring enough
for your own fame. I wish I was in every way more worthy
of your good opinion. Farewell. How pleasantly Mrs. Hooker
and you must rest from one of your many labours. . . .

* This refers to an unsolicited   towards the expenses of Sir J.
grant by the Colonial Government   Hooker's ' Flora of Tasmania.'

Again farewell : I have written a wonderfully long letter Adios, and God bless you.

My dear Hooker, ever yours,

C. DARWIN.

P.S.—I have just looked over my rambling letter ; I see that I have not at all expressed my strong admiration at the amount of scientific work, in so many branches, which you have effected. It is really grand. You have a right to rest on your oars ; or even to say, if it so pleases you, that "your meridian is past ;" but well assured do I feel that the day of your reputation and general recognition has only just begun to dawn.

[In September, 1854, his Cirripede work was practically finished, and he wrote to Sir J. Hooker :

" I have been frittering away my time for the last several weeks in a wearisome manner, partly idleness, and odds and ends, and sending ten thousand Barnacles out of the house all over the world. But I shall now in a day or two begin to look over my old notes on species. What a deal I shall have to discuss with you ; I shall have to look sharp that I do not 'progress' into one of the greatest bores in life, to the few like you with lots of knowledge."]

END OF VOL. I.

Printed in the United States
By Bookmasters